AF844241

19599

OXFORD GEOGRAPHICAL AND
ENVIRONMENTAL STUDIES

Editors: Gordon Clark, Andrew Goudie, and Ceri Peach

SOCIAL POWER AND THE URBANIZATION OF WATER

Editorial Advisory Board

Professor Kay Anderson (United Kingdom)
Professor Felix Driver (United Kingdom)
Professor Rita Gardner (United Kingdom)
Professor Avijit Gupta (United Kingdom)
Professor Christian Kesteloot (Belgium)
Professor David Thomas (United Kingdom)
Professor B. L. Turner II (USA)
Professor Michael Watts (USA)
Professor James Wescoat (USA)

ALSO PUBLISHED BY
OXFORD UNIVERSITY PRESS
IN THE OXFORD GEOGRAPHICAL AND
ENVIRONMENTAL STUDIES SERIES

The Globalized City
Economic Restructuring and Social Polarization in European Cities
Edited by Frank Moulaert, Arantxa Rodriguez, and Erik Swyngedouw

Of States and Cities
The Partitioning of Urban Space
Edited by Peter Marcuse and Ronald van Kempen

Globalization and Integrated Area Development in European Cities
Frank Moulaert

Globalization and Urban Change
Capital, Culture, and Pacific Rim Mega-Projects
Kris Olds

Sustainable Livelihoods in Kalahari Environments
Edited by Deborah Sporton and David S. G. Thomas

Conflict, Consensus, and rationality in Environmental Planning
An Institutional Discourse Approach
Yvonne Rydin

An Uncooperative Commodity
Privatizing Water in England and Wales
Karen J. Bakker

Manufacturing Culture
The Institutional Geography of Industrial Practice
Meric S. Gertler

Thailand at the Margins
Internationalization of the State and the Transformation of Labour
Jim Glassman

Social Power and the Urbanization of Water

Flows of Power

Erik Swyngedouw

OXFORD
UNIVERSITY PRESS

OXFORD
UNIVERSITY PRESS

Great Clarendon Street, Oxford OX2 6DP

Oxford University Press is a department of the University of Oxford.
It furthers the University's objective of excellence in research, scholarship, and education by publishing worldwide in

Oxford New York

Auckland Bangkok Buenos Aires Cape Town Chennai
Dar es Salaam Delhi Hong Kong Istanbul Karachi Kolkata
Kuala Lumpur Madrid Melbourne Mexico City Mumbai Nairobi
São Paulo Shanghai Taipei Tokyo Toronto

Oxford is a registered trade mark of Oxford University Press in the UK and in certain other countries

Published in the United States
by Oxford University Press Inc., New York

© Erik Swyngedouw 2004

The moral rights of the author have been asserted
Database right Oxford University Press (maker)

First published 2004

All rights reserved. No part of this publication may be reproduced, stored in a retrieval system, or transmitted, in any form or by any means, without the prior permission in writing of Oxford University Press, or as expressly permitted by law, or under terms agreed with the appropriate reprographics rights organization. Enquiries concerning reproduction outside the scope of the above should be sent to the Rights Department, Oxford University Press, at the address above

You must not circulate this book in any other binding or cover and you must impose this same condition on any acquirer

British Library Cataloguing in Publication Data
Data available

Library of Congress Cataloging in Publication Data
Data available

ISBN 0-19-823391-4

1 3 5 7 9 10 8 6 4 2

Typeset by SNP Best-set Typesetter Ltd., Hong Kong
Printed in Great Britain
on acid-free paper by
Biddles Ltd, King's Lynn

For Eva, Nikolaas, and Arno

EDITORS' PREFACE

Geography and environmental studies are two closely related and burgeoning fields of academic enquiry. Both have grown rapidly over the past few decades. At once catholic in its approach and yet strongly committed to a comprehensive understanding of the world, geography has focused upon the interaction between global and local phenomena. Environmental studies, on the other hand, have shared with the discipline of geography an engagement with different disciplines, addressing wide-ranging and significant environmental issues in the scientific community and the policy community. From the analysis of climate change and physical environmental processes to the cultural dislocations of post-modernism across the landscape, these two fields of enquiry have been at the forefront of attempts to comprehend transformations taking place in the world, manifesting themselves at a variety of interrelated spatial scales.

The 'Oxford Geographical and Environmental Studies' series aims to reflect this diversity and engagement. Our goal is to publish the best and original research in the two related fields and, in doing so, demonstrate the significance of geographical and environmental perspectives for understanding the contemporary world. As a consequence, our scope is deliberately international and ranges widely in terms of topics, approaches, and methodologies. Authors are welcome from all corners of the globe. We hope the series will assist in redefining the frontiers of knowledge and build bridges within the fields of geography and environmental studies. We hope also that it will cement links with issues and approaches that have originated outside the strict confines of these disciplines. In doing so, our publications contribute to the frontiers of research and knowledge while representing the fruits of particular and diverse scholarly traditions.

Gordon L. Clark
Andrew Goudie
Ceri Peach

Here is no water but only rock
Rock and no water and the sandy road
The road winding above among the mountains
Which are mountains of rock without water
If there were water we should stop and drink
Amongst the rock one cannot stop or think
Sweat is dry and feet are in the sand
If there were only water amongst the rock
Dead mountain mouth of carious teeth that cannot spit
Here one can neither stand nor lie nor sit
There is not even silence in the mountains
But dry sterile thunder without rain
There is not even solitude in the mountains
But red sullen faces sneer and snarl
From doors of mudcracked houses
If there were water

From: T. S. Eliot, The Waste Land

Urban and rural landscapes . . . are not two places but one. They created each other, they transformed each other's environments and economies, and they now depend on each other for survival . . . We all live in the city. We all live in the country. Both are second nature to us.

Cronon 1991: 384–5

PREFACE

The day after I first arrived in Guayaquil in mid-1992, a friend showed me around town. I was given the classic geographer's tour of the city, Ecuador's largest, located on the Pacific coast of this Andean country. The tour ended in the late afternoon on a hill on the outskirts of the central part of town. It is the kind of hill often favoured by geographers and planners to take visitors for a bird's-eye view of a city. Such a perspective helps to chart a panoramic view and 'explain' the city. This male gaze par excellence provides the illusion that it is possible to 'put the city in your pocket' as a woman friend once put it. I was tired and thirsty after a long day filled with stories and visits, and saturated with new smells, sights, and impressions.

While my friend kept pointing out the landmarks that dotted the urban landscape and invited attention, my mind and eyes wandered off to observe the intense movements of people and trucks at the foot of the hill. Large blue trucks drove on and off, while dozens of apparently similar vehicles flowed back and forth over the dusty road. This hustle and bustle suggested busy economic activity, more so than anywhere else in the city that I had just visited. I interrupted my friend's story and asked him about this strange traffic. He glanced down, commented 'Oh, they are selling water', and continued with the story that I had impolitely interrupted. I pondered the idea for a moment: 'selling water'. I also pondered the tone of self-evidence with which my friend had uttered these words. It appeared to suggest that my question was rather naïve, implying a blissful ignorance of the realities of urban life in Guayaquil. Of course they were selling water. What else would a bunch of blue trucks at the foot of a hill on the outskirts of Guayaquil do? I sat down and waited until my friend had finished his story, while watching the movement of the trucks with growing interest. Who were they selling to? Where did the water come from? What kind of water? For whose city?

I finally managed to catch my friend's attention again, and he explained. 'Well,' he said, 'you remember this afternoon we visited the informal settlements in Guasmo and Suburbio, and in Mapasingue and Barrio Popular. The people living there—over 600,000 of them in a city of roughly two million—have no piped potable water, not even standpipes. The trucks you see down there drive to these settlements and sell water door-to-door, like ice-cream. The water is actually very expensive. They pay about 450 sucre (US$0.30) to fill up a 55-gallon tank. In fact, water is one of the most serious problems in this city, together with housing, transport, and crime. These trucks are privately owned, and operate in a semi-legal framework. They buy water from the publicly owned municipal water company at a highly subsidized rate (70 sucre/1,000 litres) and sell it on. I don't have to tell you that this quasi-informal economy is very lucrative. Of course, for the people starving of thirst in the informal settlements the water vendors are both essential for their survival and considered to be thieves and crooks. There are continual tensions and little skirmishes between residents and water vendors. The relationship is rather tense. Moreover, the water system in Guayaquil is notoriously unreliable. Where you and I live (in the centre), we often don't have water either. It is all quite a mess.'

I listened to the story with growing amazement. It was hot. I was sweating all over. The smell I gave off must have been rather unpleasant. I longed for a shower and a cold drink. My gaze moved back to the panorama of the city as I tried to imagine it without water. The city began to disappear and the image of a desert, of a dry and hot wasteland began to creep into my imagination. A place without people, without water, without life. In the far distance, the mighty Guayas River flowed by. Strange. Millions of gallons of water flow through the city, yet thousands of little struggles are waged daily, by tens of thousands of people, for a bit of expensive, more or less potable, water. Of course, earlier that day I had also seen the gated communities of the upper classes, with their swimming pools, irrigated gardens, and lavish fountains decorating the entry squares of the highly protected and privately policed enclaves.

For a few years, I had been reading and thinking about politics, economics, the city; and about social power, exclusion, and revolt. My 'green' friends kept insisting that nature and the environment needed to be taken seriously as well. Perhaps they were right. What if we started thinking about the city, nature, and social power? What, if any, was the relationship between urban ecology and politics, between empowerment and disempowerment and the flow of water? What was hidden behind the H_2O that was trucked around this city? What would such an excavation of the flow of urban water tell me about the city, its people and the mechanisms of political, economic, and cultural domination? I wondered, but I also knew then that a practice and a story was hidden somewhere in that flow of water; a practice and a story of flows of liquid power. This book is the result of the search for this story.

Erik Swyngedouw

Oxford
1 July 2003

ACKNOWLEDGEMENTS

The origins of this book date back to some time in 1986, when the Catholic University of Guayaquil in Ecuador approached the Institute for Urban and Regional Planning of the University of Leuven, Belgium, where I was a researcher at the time, to seek help for establishing an institute for urban and regional planning at their university. Within a few years, a major research and institution building exercise was launched, financially supported by the Belgian Ministry of Development (ABOS) and the Flemish Interuniversity Research Council (VLIR). Although I had moved to Oxford in 1988, the project's director, Professor Louis Albrechts of the University of Leuven, invited me to continue to be involved in this Ecuadorian venture. Between 1988 and 1994, I spent more than a year in Ecuador at the newly established Instituto de Planificación Urbano y Regional (IPUR), undertaking the field research that would eventually lead to this book. I am grateful to St Peter's College and the School of Geography for granting me the sabbatical leave to undertake this research. I continued my work with further shorter visits, mainly funded by Oxford University's Hayter Fund.

In Guayaquil, I had the good fortune to work with a great team of Ecuadorean and Belgian academics. All of them have been instrumental in shaping the analysis presented in the next pages. In addition to Louis Albrechts, who has been an inspiring mentor over the years, Andrew Bovarnick, Galo Chiriboga, José Delgado, Piet Deseure, Luis Gomez, Carlos Leon, Jef Marien, Joris and Hilde Scheers, Gaetan Villavicencio and a supporting network of local friends have been instrumental in making this book come to fruition. Greet Remans was a loving companion and comrade during many of these wonderful years. Annie Collaer kept us all informed and organized with her great organizational talents. With the support of the Flemish International Centre (VIC) and with the friends from the Federación de Barrios Suburbanos (FEDEBAS) in Guayaquil, an alternative water supply and distribution system was set up in some of the informal settlements of Guayaquil. Making this project possible was for me a small, but significant, way of trying to make our research socially meaningful and politically relevant.

Of course, the theoretical framework that laid the foundations for the analysis presented in this book was developed over the years in the context of the stimulating and exciting debates, arguments, and collaborative work I enjoyed in Oxford and which have helped to shape and sharpen the arguments presented here. I owe a considerable debt to my friends, colleagues, and students in the School of Geography and the Environment and in St Peter's College. In particular, Simon Addison, Guy Baeten, Karen Bakker, Jessica Budd, Esteban Castro, Kim Hammond, David Harvey, Maria Kaïka, Alex Loftus, Ben Page, and Judith Tsouvalis not only provided an intellectually stimulating environment, but also brought the fun, love, pleasure, and enjoyment that is so often absent from the dim corridors of academic institutions. David Dodman's editorial work was meticulous and detailed. Ailsa Allen has been great as usual producing all the graphs and cartographic work.

This book is dedicated to my children Eva, Nikolaas, and Arno. Eva and Nikolaas remember their stay in Guayaquil with great fondness. I am sure Arno will one day also visit this beautiful country, Ecuador on which he has just completed his school project. My work has cost them dearly in terms of time I did not spend with them, but rather with the people of Guayaquil or sitting behind my computer. I can only hope that one day they will understand and forgive me for the time stolen from them.

CONTENTS

LIST OF PLATES

LIST OF FIGURES

LIST OF TABLES

Introduction: The Power of Water

Water is indispensable 'stuff' for maintaining the metabolism, not only of our human bodies, but also of the wider social fabric. The very sustainability of cities and the practices of everyday life that constitute 'the urban' are predicated upon and conditioned by the supply, circulation, and elimination of water. The complex web of the 'Metabolisms of Cities' (Wolman 1965: 179) relies on the perpetual circulation of water into, through, and out of the city. Without an uninterrupted flow of water, the maelstrom of city life and the mesmerizing collage of interwoven practices that constitute the very essence of urbanity are hard to imagine. It is difficult, if not impossible, for most of us to even think about living without water for drinking, washing, bathing, cooking, or cleaning for more than a few hours. Indeed, like food, water is both a biological necessity and a key economic commodity, as well as being the source of an intricate and rich cultural and symbolic power (see Bachelard 1942). But while the supply of food, clothing, and durable goods can be handled through local, decentralized, individual initiative, the supply of water is routinely—although by no means necessarily or exclusively—organized by means of large bureaucratic and engineering control systems, collective intervention and action, and centralized decision-making systems (see Wittfogel 1957; Worster 1985; Lorrain 1997; Donahue and Johnston 1998). Such centralized and hierarchical systems, whether privately or publicly owned, enable monopoly control and, given the commodity character of water, permit the extraction of monopoly profits in addition to the powerful social and political control that goes with monopolistic control over vital goods. Contrary to the rural realm where—at least under non-arid conditions—water of a reasonable quality is easily and often readily available, urban water supply and access relies on the perpetual transformation, mastering, and harnessing of 'natural' water. Urban water is necessarily transformed, 'metabolized' water, not only in terms of its physico-chemical characteristics, but also in terms of its social characteristics and its symbolic and cultural meanings. In capitalist cities, or at least in cities where market relations are the dominant form of exchange, this circulation of water is also an integral part of the circulation of money and capital. Nature's water is captured, pumped, purified, chemically adjusted, piped, bought and sold, regulated, used by households, agriculture and industry, transformed into electricity, biochemically metabolized by plants, animals and humans, integrated in public displays like fountains, often turned into sewage, eventually returned to 'nature'. As with other urban goods and services, water circulation is part and parcel of the political economy of power that gives structure and coherence to the urban fabric. Indeed, the water/money nexus combined with H_2O's essential life-giving and life-sustaining use-value inserts water and the hydrosocial cycle into the power relationships of everyday life and makes it subject to intense social struggle along class, gender, and ethnic cleavages for access and/or control. Mechanisms of access to and exclusion from water lay bare political economic power relationships and positions of social and cultural power, particularly in cities that lack adequate water supply systems or in environments characterized by heavily contested water usage. The circulation of water combines political and economic power at the international, national, regional, and local levels with a social and economic struggle for the control over and appropriation of water. Both public and private agents are deeply implicated in this struggle for the command over water and for power.

The flow of water and the flow of money and power are, consequently, materially linked. In a variety of ways, Worster (1985), Reisner (1986), Davis (1990), and Hundley (1992) have shown how watering California in general, and Los Angeles in particular, has been a tumultuous and conflict-ridden process driven by relations of political power, economic control, and territorial conflict. In the same way, I suggest that the power/money/water nexus can be introduced as a conceptual triad, which lays bare the political economy of the urban fabric and the functioning of mechanisms of domination and subordination within the urban arena. Just as the investigation of the circulation of money and capital illustrates the functioning of capitalism as an economic system (see Harvey 1981; 1982), I aim to demonstrate that the circulation of water—as a physical and social process—brings to light wider political economic, social, and ecological processes. In turn, this will permit a better understanding of the political ecological processes that shape urbanization. Indeed, controlling the flow of water implies controlling the city, as without the uninterrupted flowing of water, the city's metabolism would come to a halt. The metaphorical and material streams of power that give Guayaquil, or any other city, its city structure can be unravelled and reconstructed, I hold, through excavating the political economic relations through which water is brought into, circulated through, and taken out of the city. And this is exactly the task set out for this book.

The particular irony evident in Guayaquil, is that billions of litres of water flow through the city centre as the Rivers Daule and Babahoyo come together to form the mighty Guayas stream, while almost half the city dwellers do not have access to adequate and reliable potable water supplies and the entire city suffers from chronic water shortages. The sewage system, the other half of the circulatory water system, is on the verge of total collapse. For the 'invasiones', land invaded and occupied by rural migrants and the rapidly expanding urban underclass, the irony takes even more grotesque forms. The further consolidation and expansion of invasion settlements in the Guayas river estuary is organized through a detailed division of labour, often concerned with controlling and 'engineering' estuary water (landfill, elevated housing and pathway construction, simple dams, etc.) while, once new sites are occupied, the new city dwellers suffer from chronic potable water supply problems and lack of sanitary services. Despite being surrounded by saline and polluted estuary water, and being inundated during the rainy season, they never have access to adequate drinking water. The absence of water and the exclusionary practices through which the urban water supply system is organized tell a story of urban deprivation, disempowerment, and repressive social mechanisms that turn slum life into the antithesis of modern urban life.

This book seeks to document and analyse the power of water in the context of Guayaquil's urbanization process and to suggest strategies for an emancipatory and non-exclusive production, conduction, and distribution of urban water. In the first three chapters of the book, I attempt to chart the political ecological perspectives that have inspired the research on Guayaquil's urbanization process. I shall start with outlining how water captures and fuses together physical and social processes. This will set the scene for Chapter 2, where I explore the thorny relationship between nature, society, and water as they become welded together in the city through the urbanization process. The historical geography of urban water control will be briefly recapitulated to highlight the social constructedness of water use and mastering, and the material and symbolic power mechanisms that are inscribed in the way the urbanization of water has unfolded. The third chapter, then, switches the vantage point to the Latin American city and to Guayaquil, in particular, and charts the oppressive and exclusive processes that produce highly uneven and deeply problematic access to water, and in particular potable water, to many urban residents. I shall explore the flows of power and the mechanisms of participation and exclusion that describe the rituals of everyday urban life as they are inscribed in the metabolic circulation of urban water.

The second part of the book will delve into the political-ecological dynamics through which the contemporary urban waterscape and hydrosocial cycle in Guayaquil became constituted. The city's waterscape is indeed a manufactured landscape, one that is wrought, historically and geographically, from a mesmerizing mixture of local, regional, national, and international socio-economic and political-ecological processes and struggles. Chapters 4 and 5 undertake this history of the urban water networks, and reconstruct how Guayaquil's twentieth-century history became etched into the technical, social, and ecological structures of the water system. This history and current geography of the city will be written from the perspective of the necessity to control and harness water flow into and around the city. The socio-economic and political-geographic power relationships determining access to or exclusion from water will be analysed in the context of Guayaquil's urbanization process. In addition, I shall explore how these practices vindicate social and economic power relationships at the local, national, and international level.

In the subsequent part the 'Water Mandarins', which organize and control the production, conduction, and distribution of urban water in Guayaquil, will be charted with a focus on their internal and external relations. This will include an analysis of the relationship between external funding agencies (the World Bank and others), national government, and the local and recently privatized water company (Chapter 6). In addition, infrastructure and investment planning, price mechanisms, and control structures will be explored in the light of the disempowering mechanisms of the existing water system. In Chapter 7, I shall explore the relations between the water company and the 'water speculators', the 'informal' system of water distribution by a series of private water vendors ('tanqueros') that serve the suburban areas by means of tankers.

In the final section of the book, the struggles for water power will be documented. In Chapter 8, the strategies of the water company, the 'tanqueros', and the local communities will demonstrate how control over and access to water is highly contested terrain. The flow of money from the community to the state, the private sector, and the 'water speculators', and the consequent draining of resources will be detailed. Attention will be paid to both informal struggles, political clientelist strategies and to 'water violence' in the quest for control over water. These struggles exemplify the dynamics of the Guayaquileño urban political economy and highlight the mechanisms of domination/subordination and participation/exclusion in the context of peripheral urbanization processes. Attention will also be paid to 'people power', to the weapons deployed by the weak, and the ingenious mechanisms mobilized by individuals and social groups alike to secure access to at least some of the available water. The section will conclude with a discussion of the struggles over access to water in the practices of everyday urban life. In the concluding chapter, strategic issues related to the possibilities for an emancipatory and empowering development will be explored. Political, institutional, and technological alternatives enabling a more equitable water supply and distribution system and permitting local residents to exercise 'the right to the city' (and its water) will be outlined.

In short, in what follows, I aim to reconstruct the political, social, and economic conduits through which water flows and to identify how power relations infuse the metabolic transformation of water as it becomes urban. These flows of water that are simultaneously physical and social carry in their currents the embodiment of myriad social struggles and conflicts. The exploration of these flows narrates stories about the city's structure and development. Yet these flows also carry the potential for an improved, more just, and more equitable right to the city and its water. The flows of power that are captured by urban water circulation also suggest how the question of urban sustainability is not just about achieving sound ecological and environmental conditions, but first and foremost about a social struggle for access and control; a struggle not just for the right to water, but for the right to the city itself.

PART I

Flows of Power: Nature, Power, and the City

1

Hybrid Waters: On Water, Nature, and Society

> We have before us, here and now, a whole. It is both the condition for production and the product of action itself, the place for mankind and the object of its pleasure: the earth.
>
> Lefebvre 1995: 133

> . . . a thing cannot be understood or even talked about independently of the relations it has with other things. For example, resources can be defined only in relationship to the *mode of production* which seeks to make use of them and which simultaneously 'produces' them through both the physical and mental activity of the users. There is, therefore, no such thing as a resource in abstract or a resource which exist as a 'thing in itself'.
>
> Harvey 1980: 212

> A cyborg is a cybernetic organism, a hybrid of machine and organism, a creature of social reality as well as a creature of fiction.
>
> Haraway 1991: 149

> The two extremes, local and global, are much less interesting than the intermediary arrangements that we are calling networks . . . Is it our fault if the networks are simultaneously real, like nature, narrated, like discourse, and collective, like society?
>
> Latour 1993: 122, 126

1.1 Water and the urbanization of nature

1.1.1 Critical waters

In recent years, we have become increasingly aware of the importance of water as a critical good, and questions of water supply, access, and management, both in quantitative and qualitative terms, have become key issues (Gleick 1993; Postel 1992; Stauffer 1998). The proliferating commodification and privatization of water management systems; the combination of Global Environmental Change with increased demands from cities, agriculture, and industry for reasonably clean water; the inadequate access of almost a billion people on the planet to clean water (over half of whom live in large urban centres); the proliferating geopolitical struggle over the control of river basins; the popular resistance against the construction of new megadams; the political struggles around water privatization projects; and many other issues; have brought water politics to the foreground of national and international agendas (Shiklomanov 1990; 1997; Herrington 1996; Roy 2001).

In the twentieth century, water scarcity was seen as a problem primarily affecting developing societies (Anton 1993). However, at the turn of the new century, water problems are becoming increasingly globalized. In Europe, the area bordering the Mediterranean, notably Spain, southern Italy, and Greece, is arguably the location in which the water crisis has become most acute, both in quantitative and qualitative terms (Batisse and Gernon 1989; Margat 1992; Swyngedouw 1996*a*). However, northern European countries, such as the UK, Belgium, and France, have also seen increasing problems with water supply, water management, and water control (Haughton 1996), while transitional societies in eastern Europe are faced with mounting water supply problems (Thomas and Howlett 1993). The Yorkshire drought in England, for example, or the Walloon/Flemish dispute over water rights are illuminating examples of the intensifying conflict that surrounds water issues (Bakker 1999). Cities in the global South and the global North alike are suffering from a deterioration in their water supply infrastructure and in their environmental and social conditions in general (Lorrain 1995; Brockerhoff and Brennan 1998). Up to 50% of urban residents in the developing world's megacities have no easy access to reasonably clean and affordable water. The myriad socio-environmental problems associated with deficient water supply conditions threaten urban sustainability, social cohesion, and, most disturbingly, the livelihoods of millions of people (Niemczynowicz 1991). It is not surprising, then, to find that issues of water have become highly contested. Political conflict, ecological problems, and social tensions multiply as the competition for access to water intensifies (Worster 1985; Hundley 1992; Shiva 2002). Yet, cities are becoming increasingly thirsty (Cans 1994). This book will concentrate on the thorny relationship between the urbanization process and socio-ecological conditions. In the process, it will argue that urbanization is primarily a particular socio-spatial process of metabolizing nature, of urbanizing the environment.

Urban water issues have traditionally been approached from a predominantly engineering, economic, or managerial approach, with precious little attention paid to the central role of social and political questions (Goubert 1989). The social risk associated with growing water problems as manifestations of wider socio-ecological and political ecological changes have been even less scrutinized. The problematic water supply and access conditions in many of the cities in the Global South—cities as varied as Jakarta, Mexico City, Lagos, Cochabamba, or Guayaquil—testify to the growing risk and associated social and political tensions in this domain. In light of mounting environmental concerns (global climate change, pollution, soil degradation, etc.), environmental risks are viewed as becoming increasingly central to political and social issues, debates, and approaches (see Beck 1992; 1995). In light of real or perceived risks of water crises, a review of the way in which the hydrological cycle, water management, water politics, and water economics are understood and theorized is long overdue.

It is in many ways astonishing that in the ballooning literature on the environment and among the innumerable environmental social movements, the city often figures in a rather marginal or, worse, an antithetical manner. Even more surprising is the almost complete absence of a serious engagement with the environmental problematic in the prolific literature on the city.[1] At a time when the world is rapidly approaching a situation in which more than half of its population dwells in large cities, the environmental question is generally often circumscribed to either rural or threatened 'natural' environments or to 'global' problems. Yet, the urbanization process is central to the momentous environmental changes and alleged problems that have inspired the emergence of environmental issues on the political agenda.

1.1.2 The urbanization of nature

In *Justice, Nature and the Geography of Difference*, David Harvey (1996) insists that there is nothing particularly 'unnatural' about New York City or any other city. Cities are dense networks of interwoven socio-spatial processes that are simultaneously human, material, natural, discursive, cultural, and organic. The myriad transformations and metabolisms that support and maintain urban life, such as water, food, computers, and movies always involve infinitely interconnected environmental *and* social processes (Swyngedouw 1999). Imagine, for example, standing on a busy street corner of any city in the world and considering the socio-environmental metabolic relations that come together in this global/local place. Smells, tastes, and bodies from all nooks and crannies of the world are floating by, consumed, displayed, narrated, visualized, and transformed. Shops and restaurants play to the tune of eco-sensitive shopping and the multi-billion pound eco-industry while competing with McDonalds' burgers and Dunkin' Donuts. The sounds of world music vibrate from music shops while people, spices, clothes, foodstuffs, and materials from all over the planet whirl by. The neon lights are fed by the processes of nuclear fission, coal, or gas burning in far-off power plants, while the cars consume fuels from oil deposits and pump CO_2 into the air, affecting forests and climates around the globe. These disparate processes trace the global geographic mappings that flow

[1] With some exceptions, such as: Davis (1990; 1995; 1998), Cronon (1991), Keil (1994; 1995; 1998), Keil and Desfor (1996), Gandy, (1996; 1999; 2002), Harvey (1996), Swyngedouw (1996*b*; 1997), Swyngedouw and Kaïka (2000); Kaïka and Swyngedouw (2000).

through the urban landscape and 'produce' cities as palimpsests of densely layered bodily, local, national, and global—but geographically depressingly uneven—socio-ecological processes. This intermingling of things material and things symbolic produces a particular socio-environmental milieu that welds nature, society, and the city together in a deeply heterogeneous, conflicting, and often disturbing whole (Swyngedouw 1996*b*). The socio-ecological footprint of the city has become global. There is no longer an outside or limit to the city, as the urban process harbours social and ecological processes that are embedded in dense and multilayered networks of local, regional, national, and global connections.

In the emerging literature on 'the sustainable city', little attention has thus far been paid to the urban as a process of socio-ecological *change*,[2] while discussions about global environmental problems and the possibilities for a 'sustainable' future customarily ignore the urban origin of many of the problems. Of course, 'environmental' issues have been central to urban change and urban politics for at least a century if not more. Visionaries of all sorts lamented the 'unsustainable' character of early modern cities and proposed solutions and plans that would remedy the antinomies of urban life and produce a healthy 'wholesome' urban living. As Raymond Williams pointed out in *The Country and the City* (1985 (1973)), the transformation of nature and the social relations inscribed therein are inextricably connected to the process of urbanization. The dialectic of the environment and urbanization consolidates a particular set of social relations through 'an ecological transformation which requires the reproduction of those relations in order to sustain it' (Harvey 1996: 94). These socio-environmental changes result in the continuous production of new 'natures', of new urban social and physical environmental conditions. All of these processes occur in the realms of power in which social actors strive to defend and create their own environments in a context of class, ethnic, racial and/or gender conflicts and power struggles. Of course, under capitalism, the commodity relation veils and hides the multiple socio-ecological processes of domination/subordination and exploitation/repression that feed the capitalist urbanization process and turn the city into a metabolic socio-environmental process that stretches from the immediate environment to the remotest corners of the globe. Indeed, the apparently self-evident commodification of nature that fundamentally underpins a market-based society not only obscures the social relations of power inscribed therein, but also permits the disconnection of the perpetual flows of transformed and commodified nature from its inevitable foundation, i.e. the transformation of nature (Katz 1998). In sum, the environment of the city (both social and physical) is the result of a historical geographical process of the urbanization of nature (Swyngedouw and Kaïka 2000).

[2] Among those who address the issue are Blowers (1993) and Haughton and Hunter (1994), or, for a more critical perspective, Burgess, Carmona, and Kolstee (1997), Baeten (2000) and Gandy (2002).

Although Henri Lefebvre does not address the environment of the city directly, he does remind us of what the urban really is, i.e. something akin to a vast and variegated whirlpool replete with all the ambivalence of a space full of opportunity, playfulness, and liberating potential, while being entwined with spaces of oppression, exclusion, and marginalization (Lefebvre 1991 (1974)). Cities seem to hold the promise of emancipation and freedom whilst skilfully mastering the whip of repression and domination (Merrifield and Swyngedouw 1996). Ironically, the relations of domination and power that infuse urban practices and which are contested in innumerable ways help to create the differentiated environments that give cities their sweeping vitality. At the same time, these forms of resistance and subversion of dominant values tend to perpetuate the conservative imagery of cities as places of chaos, social, and environmental disintegration, and moral decay. Perpetual change and an ever-shifting mosaic of environmentally and socio-culturally distinct urban ecologies—varying from the manufactured and manicured landscaped gardens of gated communities and high-technology campuses to the ecological war-zones of depressed neighbourhoods with lead-painted walls, asbestos-covered ceilings, waste dumps, and pollutant-infested areas—still shape the choreography of a capitalist urbanization process. Environmental ideologies, practices, and projects are part and parcel of this dialectical process of the urbanization of nature. Needless to say, the above constructionist perspective considers the process of urbanization to be an integral part of the production of new environments and new natures, which sees both nature and society as fundamentally combined historically geographical production processes (see, among others, Smith 1984; 1996; 1998; Castree 1995). This perspective has major consequences for political strategy. As Lewontin insists:

> [T]he constructionist view . . . is of some consequence to human action. A rational environmental movement cannot be built on the demand to save the environment, which, in any case, does not exist. . . . Remaking the world is the universal property of living organisms and is inextricably bound up with their nature. Rather, we must decide what kind of world we want to live in and then try to manage the process of change as best we can approximate it. (Lewontin 1997: 137–8)

In this sense, there is no such thing as an unsustainable city in general, but rather there are a series of urban and environmental processes that negatively affect some social groups while benefiting others. A just urban socio-environmental perspective, therefore, always needs to consider the question of who gains and who pays and to ask serious questions about the multiple power relations through which deeply unjust socio-environmental conditions are produced and maintained. This requires sensitivity to the political ecology of urbanization rather than invoking particular ideologies and views about the qualities that are assumed to be inherent in nature itself. Before we can embark on outlining the dimensions of an urban political ecological enquiry, we need to consider the matter of nature in greater detail, in particular in light of the accelerating process by which nature become urbanized through the deepening metabolic interactions between social and ecological processes.

1.2 The question of nature: hybrid worlds

There remains nothing, in culture or nature, which has not been transformed, and polluted, according to the means and interests of modern industry. (Guy Debord 1990: 10)

In fact, 'nature' is merely the uncoded category that modernists oppose to 'culture', in the same way that, prior to feminism, 'man' was the uncoded category opposed to 'woman'. By coding the category of 'natural object', anthropological science loses the former nature/culture dichotomy. Here, there is an obvious link with feminism. Nothing more can be done with nature than with the older notion of man. (Bruno Latour 1998: 238)

Early in 1998 (*Le Monde*, 17 January), controversy arose in the Paris region about IBM's continued tapping of ancient underground aquifers. The company's manufacturing processes require large volumes of water of the highest purity to cleanse the micropores on chips. Environmentalists seeking to protect historical 'natural waters' were outraged; the water company, Lyonnaise des Eaux, was worried about the potential loss of water and, consequently, of future dividends; while the state at a variety of scales was caught up in the myriad tensions ensuing from this: protection of the natural environment versus economic priorities, the competing claims of different companies, etc. The ancient underground waters fused with politics, economics, and culture in intricate ways.

Later the same year, the Southeast Asian financial bubble imploded. Global capital moved spasmodically from place to place, leaving cities like Jakarta with social and physical wastelands in which dozens of unfinished skyscrapers are dotted over the landscape while thousands of unemployed children, women, and men roam the streets in search of survival. In the meantime, El Niño's global dynamic was wreaking havoc in the region with its climatic disturbances. Concrete buildings that had once promised continued capital accumulation for Indonesia now held nothing more than puddles of stagnant water providing breeding grounds for mosquitoes. Malaria and dengue fever suddenly joined with unemployment and social and political mayhem in shaping Jakarta's cityscape. Global capital fused with global climate, with local power struggles, and with socio-ecological conditions to reshape Jakarta's social ecology in profound, radical, and deeply troubling ways.

In 1997, the scientific community was thrilled and the wider public shocked when Scottish researchers revealed that they had been able for the first time to clone a higher organism. The cloned sheep, appropriately called 'Dolly',[3] outraged environmentalists, initiated a moral debate about the rights of humans to tamper with and reconstruct nature, excited the bio-technology venture capitalists who imagined a burgeoning multi-billion pound new industry, and incensed feminists who considered 'Dolly' to be a soulmate subjected to the whims and desires of a patriarchical, domineering, and manipulative male order. Meanwhile, the scientists congratulated themselves on their breakthrough in disentangling and commanding the web of life.

These are just three examples from a proliferating number of cases in which the traditional distinction between environment and society, between nature and culture becomes blurred, ambiguous, and problematic. They also capture current arguments over the 'nature of nature'. What I wish to undertake here is a more detailed exploration of the challenges and implications arising from the examples given above, and a tentative suggestion of avenues for exploring and transforming the world in an emancipatory fashion. The above stories exemplify what is at the core of Bruno Latour's critique of the purifying rituals that have plagued modern science ever since the Enlightenment. The desire of scientists to divide the world into two separate poles, nature on the one hand and culture on the other, seems to have lost much of its explanatory and political power in an era when it is becoming increasingly apparent that things 'natural' and things 'cultural' do not exist side by side as the two opposite poles of a dialectical unity. As Latour's quote suggests, we have to abandon the categories of nature and culture altogether. In *We Have Never been Modern*, Latour (1993) argues how the Gordian knot that weaves together the 'natural' and 'social' has been cut through by the sword of the purifying rituals that became encoded in the scientific enterprise of the Enlightenment. It was precisely this unruly binarization that permitted scientists and engineers to decode some of the intricacies of parts of the world (while, of course, being totally unaware of the socially and culturally significant meanings that became scripted into their scientific explanations). More importantly, the particular knowledge of the 'purified' natural world that was generated by the practices and gazes of the scientists permitted precisely the proliferation of the hybrid 'things' mentioned above. Scientific knowledge and practices fused with physical metabolic processes to produce socio-natural and socio-technical hybrid complexes. In many ways, the separation between nature and society accelerated the formation of these socio-natural cyborgs and quasi-objects of which H_2O, 'Dolly', or Oncomouse[tm] (see Haraway 1991; 1997; Lykke 1996) have become canonical examples. Similarly, urban and regional landscapes, climate change, ozone

[3] It is, of course, not a coincidence that the cloned sheep was female and given a name usually associated with a female playmate, which combines male-oriented sexually explicit characteristics with a naivety that renders her easily subject to male fantasies and manipulations. 'Dolly' became an icon of a commodified, sexist, and manipulative academic industry.

depletion in the stratosphere and ozone overconcentration in the troposphere, El Niño and the forest fires in Indonesia, prions and BSE, the threat of perennially polluted drinking water, and risks of droughts and floods, the daily struggle many have to wage to obtain reasonably clean water all testify to the myriad ways in which the natural and the social have transgressed and continue to blur the boundaries that modern science, including geography, have tried to spin around the 'natural' and 'social' worlds. Indeed, on closer inspection, the city, water, ozone, BSE, 'Dolly', and human bodies are networks of interwoven processes that are simultaneously human and natural, real and fictional, mechanical and organic. There is nothing 'purely' social or natural about them, even less asocial or a-natural: these 'things' are natural *and* social, real *and* fictional. Society and nature, representation and being, are inseparable, integral to each other, infinitely bound up. Simultaneously, these hybrid socio-natural 'things' are full of contradictions, tensions and conflicts (Castree and MacMillan 2001).

Their very existence has a lot to say both about modernity's project of purification and about self-described 'post-modern' debates on the importance of the sign. To start with the latter, the existence of hybrids of the kind exemplified above is a constant reminder and proof of the impossibility of separating 'representation' from 'being', the sign from the signified, the discursive from the material. In terms of what this hybridization has to say about modernity's purification project, I shall argue that the way in which these socio-natural 'hybrids' encompass contradictions, tensions, and conflicts shows that the scientific endeavour of slicing the 'Gordian knot' binding the 'natural' and the 'social' together has only been accomplished at a discursive/scientific level. The separation worked at the epistemological level, that is as a way of understanding the world, and as such has indeed managed to produce knowledge. The problem with this epistemological perspective, once it became hegemonic, is that it eventually turned from a dominant epistemology to a dominant ontology, that is a strong belief that the world was actually ontologically split into things natural and things social. This translocation of epistemology into ontology was not of course without profound social, political, and cultural implications and was indeed highly relevant to the historical, social, cultural, and political background against which it happened. As Latour argues, the proliferation of 'hybrids' permits (and even necessitates) everyone (including scientists) to see the impossibility of an ontological basis for such a separation. Their very existence is proof of the flaw of such an argument. The irony, of course, is that 'hybridization' emerged precisely from the very laboratories whose fundamental purpose had been to rule out (outlaw) hybridity.

In the next section, I shall tentatively suggest a research programme to explore the proliferation of quasi-objects and cyborgs in the present world, and attempt to contribute to the formulation of an emancipatory political-ecological programme.

1.3 On hybrids and socio-nature: flow, process, and dialectics

1.3.1 The materialist legacy

Karl Marx's historical materialism was arguably the first coherent attempt to theorize the internal metabolic relationships that shape the transformations of the earth's surface. In *Grundrisse*, in *Capital* and, in particular, in *The German Ideology*, Marx insisted on the 'natural' foundations of social development:

> The first premise of all human history is, of course, the existence of living human individuals. Thus the first fact to be established is the physical organisation of these individuals and their consequent relationship to the rest of nature . . . The writing of history must always set out from these natural bases and their modification in the course of history through the action of men . . . [M]en must be in a position to live in order to be able to 'make history' . . . The first historical act is thus the production of the means to satisfy these needs, the production of material life itself. (Marx (1974 (1846): 42, 48)

This 'production' process is basically a labour process (in the widest possible sense of the word). Labouring is therefore nothing other than engaging the 'natural' physical and mental forces and capabilities of humans in a metabolic physical/material process with other human and non-human natural conditions and actors. 'Metabolism' is the central metaphor for Marx's definition of labour and for analysing the relationship between human and nature:

> Labour is, first of all, a process between man and nature, a process by which man, through his own actions, mediates, regulates, and controls the metabolism between himself and nature. He confronts the materials of nature as a force of nature. He sets in motion the natural forces which belong to his own body, his arms, legs, head, and hands, in order to appropriate the materials of nature in a form adapted to his own needs. Through this movement he acts upon external nature and changes it, and in this way he simultaneously changes his own nature. . . . [labouring] is the purposeful activity aimed at the production of use-values. It is an appropriation of what exists in nature for the requirements of man. It is the universal condition for the metabolic interaction between man and nature, the everlasting nature-imposed condition of human existence, and it is therefore independent of every form of that existence, or rather it is common to all forms of society in which human beings live. (Marx 1971 (1867): 283, 290)

For Marx, this socio-natural metabolism is the foundation of and possibility for history, a socio-environmental history through which the nature of humans and non-humans alike is transformed. To the extent that labour constitutes the universal premise for metabolic interaction with nature, the particular social relations through whom this metabolism of nature is enacted shape its very form. Clearly, any materialist approach insists that 'nature' is an integral part of the 'metabolism' of social life. Social relations operate in and through metabolizing the 'natural' environment and transform both society and nature.

Marx undoubtedly borrowed the notion of 'metabolic interaction' from von Liebig, the founding theoretician of modern agricultural chemistry. In fact, the original German word is 'stoffwechsel', which simultaneously means circulation, exchange AND transformation of material elements. As Foster (2000) argues, the notion of 'metabolism' is central to Marx's political economy and is directly implicated in the circulation of commodities and, consequently, of money:

> The economic circular flow then was closely bound up, in Marx's analysis, with the material exchange (ecological circular flow) associated with the metabolic interaction between human beings and nature. (Foster 2000: 157–8)

Under capitalist social relations, then, the metabolic production of use-values operates in and through specific control and ownership relations and in the context of the mobilization of both nature and labour to produce commodities (as forms of metabolized socio-natures) with an eye towards the realization of the embodied exchange value. The circulation of capital as value in motion is, therefore, the combined metabolic transformations of socio-natures in and through the circulation of money as capital under social relations that combine the mobilization of capital and labour power. New socio-natural forms are continuously produced as moments and things in this metabolic process (see Grundman 1991; Benton 1989; 1996; Burkett 1999; Foster 2000). While nature provides the foundation, the dynamics of social relations produce nature's and society's history. Whether we consider the production of dams, the re-engineering of rivers, the transfiguration of DNA codes, or the construction of a skyscraper, they all testify to the particular capitalist social relations through which socio-natural metabolisms are organized. Of course, the ambition of classical Marxism was wider than reconstructing the dialectics of historical socio-natural transformations and their contradictions. It also insisted on the ideological notion of 'nature' in bourgeois science and society and claimed to uncover the 'real' *Truth* through the excavation of 'underlying' socio-ecological processes (Schmidt 1971; Smith 1984; Benton 1989). As Marx insisted in *Grundrisse*:

> It is not the unity of living and active humanity with the natural, inorganic conditions of their metabolic exchange with nature, and hence their appropriation of nature, which requires explanation, or is the result of a historic process, but rather the separation between these inorganic conditions of human existence and this active existence, a separation which is completely posited only in the relation of wage labour and capital. (Marx 1973 (1858): 489)

However, by concentrating on the labour process per se, some Marxist analysis—particularly during the twentieth century—tended to replicate the very problem it meant to criticize. By relegating nature to the substratum for the unfolding of social relations, in particular labour relations, it maintained the material basis for social life while relegating 'natural processes' to a realm outside the social. Ironically, this is almost identical to the bourgeois ideology which views nature as external to society, yet universal in its functioning. Put simply, the overemphasis on the social relations under capitalism that characterized much of Marxist analysis tended to abstract away from or ignore the metabolic relation with nature and resulted in a partial blindless in twentieth-century Marxism to questions of political ecology and socio-ecological metabolisms.

Neil Smith's *Uneven Development* (1984) represented a milestone in the repositioning of nature and social-natural metabolism within the core of historical materialist analysis, by insisting that nature is an integral part of a 'process of production'. The latter concept, borrowed from Henri Lefebvre (1991 (1974)), suggests that nature itself is a historical geographical process (time/place specific), insists on the inseparability of society and nature, and maintains the unity of socio-nature as a historically produced thing. In brief, both society and nature are produced, hence malleable, transformable, and transgressive. Smith does not suggest that all non-human processes are socially produced, but argues that the idea of some sort of pristine nature ('First Nature' in Lefebvre's account) becomes increasingly problematic as historical socio-nature produces entirely new 'nature' over space and time, and the number of hybrids and quasi-objects proliferates and multiplies. Indeed, the objects and subjects of daily life have always been socio-natural, and with the process of 'modernization' have become increasingly so. Consider, for example, the socio-ecological transformations of entire ecological systems (through agriculture, for example), sand and clay metabolized into concrete buildings, or the contested production of new genomes such as Oncomouse^{tm} (Haraway 1997). Anthony Giddens (1990) suggests that in this context we have reached 'The End of Nature'. Of course, he does not imply that nature has disappeared, but rather that there is no longer anything 'out there' which has not been transformed, tainted, or metabolized by society/culture. Whereas pre-modernity was subject to the consequences of nature, modernity attacked nature by transforming it. The 'End of Nature' implies, therefore, the construction of a new nature, a nature that still hides serious consequences from humans. This is the theme Ulrich Beck (1992; 1995) elaborates. The possibility of producing 'new' nature, ranging from nuclear installations to dams, entails the proliferation of 'risk'. Risk should be understood here not in terms of hazards, but in terms of the unexpected and unknowable implications of producing new nature and the problems that individuals, social groups, states, and science face in the process. A new modernity looms around the corner, one in which tension and conflict are still rife, but one which also holds the promise of fabricating socio-nature more in tune with the desires, aspirations, and demands of humans.

In sum, the 'world' is a historical geographical process of perpetual metabolism in which 'social' and 'natural' processes combine in a historical geographical 'production process of socio-nature' whose outcome (historical nature) embodies chemical, physical, social, economic, political, and cultural

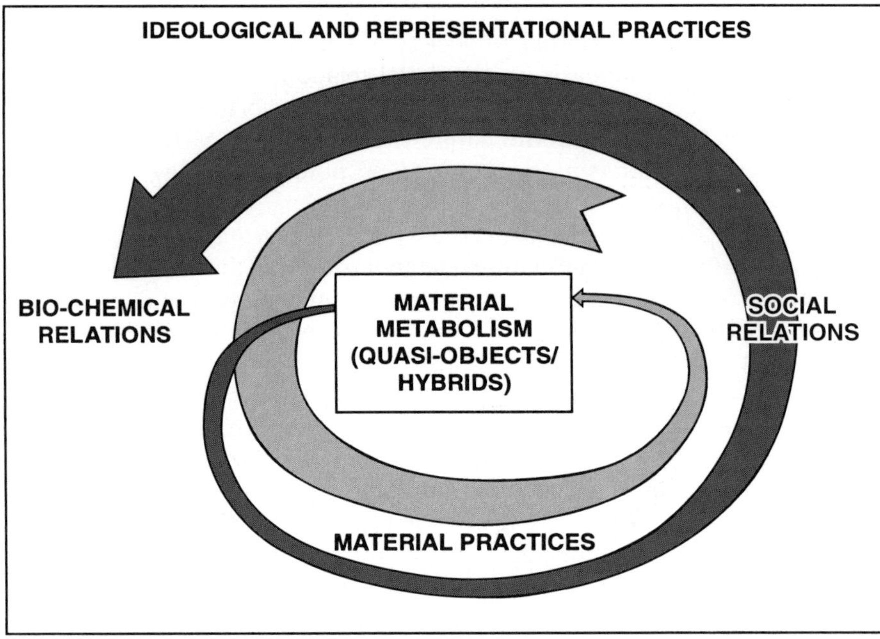

Fig. 1.1. The dialectics of the material production of socio-nature.

processes in highly contradictory but inseparable manners (Castree 2001). Every body and every thing is a cyborg, a mediator, part social, part natural, lacking discrete boundaries and internalizing the multiple contradictory relations that redefine and rework them. Take the example of urban water. Drinking tap water combines the circulation of productive, merchant and financial capital with the production of land rent and their associated class relations; the ecological transformation of hydrological complexes and the biochemical process of purification with the libidinous sensation and the physiological necessity of drinking fluids; the social regulation of access to water with images of clarity, cleanliness, health, and virginity. Although it is impossible to separate these 'concepts' and practices from each other in the flow of water, it does not take much to identify the profound social, cultural, political, and ecological forces, struggles and power relations at work in this perpetual metabolizing circulation process of flowing water as represented diagramatically in Figure 1.1.

1.3.2 The cultural critique

However 'true' the above may be, it nevertheless remains caught in a 'representational' discourse of knowledge production which denies, ignores or, at least, fails to problematize how this representation of socio-nature is itself inevitably

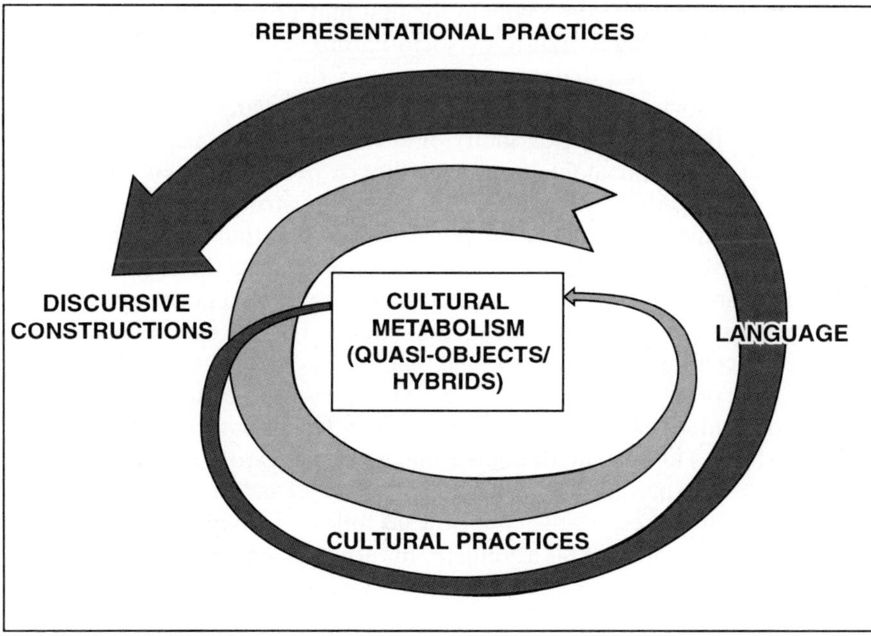

Fig. 1.2. The dialectics of the representational production of socio-nature.

caught in a web of symbolic and discursive meanings. Recent 'post-Marxist' or 'post-'modernist accounts which have challenged the possibility of constructing 'The Truth' about the world, have therefore challenged the very assumption on which the above rests. Many Western historical materialist perspectives made such truth claims their own, both in attempting to 'unveil' the ideological construction of other epistemes and in arguing for the 'real' science of historical materialism. As Castree (1995) argues, historians of science and cultural theorists alike have insisted that 'socio-nature' is not just 'out there' but becomes constructed via time/place specific modes of technological, political, and 'staged' appropriation of 'filtered facts' ('scientific' experiments or methods are a case in point); that the production of 'knowledge(s)' proceeds in and through representational systems or discursive apparatuses where reality resides in the representation, yet remains outside of it; and that the presumed correspondence of the concept with the thing is as much infused with the 'cultural' position of the representer as with the materiality of the process represented. Put simply, the representation of 'reality' previously described remains caught in the sociocultural situatedness of the times and places of representation (Whatmore 2002). Sensitivity to the constructions of representations of and discourses on 'socio-nature' are diagrammatically represented in Figure 1.2, which is apparently fundamentally at odds with Figure 1.1.

Despite the implicit claim made in the first half of this section of the possibility of constructing a 'Truth' of socio-nature via a historical materialist analysis of the internal dialectical relations of the perpetual socio-physical metabolism of socio-nature, cultural critics and historians of science (nature) not only question the very possibility of such a claim, but also, more importantly, insist on the inevitable non-neutrality or positionality of such claims (Haraway 1991; Latour 1993; 1999). In short, constructing knowledge is in itself a deeply historical, dialectical, power-full process that is infused with and embodies the very metabolisms it claims to reconstruct as the very materiality of socio-nature itself. The 'real' metabolism encircled by a political ecological episteme is itself encapsulated within and engulfed by the equally real discursive/linguistic/cultural construction of reality. Put simply, our claim of the socio-natural production process of socio-nature is itself caught in a representational discourse that produces nature/society (socio-nature) in a particular partial fashion. This insight, of course, makes our claim to 'truth' as vulnerable to 'relativist' interpretations as any other. Yet we cannot easily dismiss these post-Enlightenment criticisms, and in what follows I shall outline a possible way out of this paradox. This argument, in turn, will frame the analysis of the urbanization of water in Guayaquil.

1.4 Muddling through HYBRIDITY: flow, process, and dialectics

I believe that Henri Lefebvre's work, properly amended, can come to the rescue here. For Lefebvre (1991), capturing space or socio-nature from a dialectical and emancipatory perspective implies constructing multiple narratives that relate material, representational, and symbolic practices, each of which have a series of particular characteristics, and internalizes the dialectical relations defined by the other 'domains', but none of which can be reduced to the other. Of course, the production process of socio-nature includes both material processes (constructing edifices and manufacturing new genetic materials) as well as the proliferation of discursive and symbolic representations of nature. As Lefebvre (1991) insisted, the production of nature (space) transcends merely material conditions and processes, and becomes related to the production of discourses on nature (mainly by scientists, engineers, and the like) on the one hand and powerful images and symbols inscribed in this thing called 'nature' (virginity, a moral code, originality, 'survival of the fittest', wilderness, etc.) on the other. In short, Lefebvre's triad opens up an avenue for enquiry which insists on the 'materiality' of each of the component elements, but whose content can be approached only via the excavation of the metabolism of their becomings in which the internal relations are the signifying and producing mechanisms. In other words, Lefebvre insists on the ontological priority of process and flux, which becomes interiorized in each of the moments of the production process, but always in a fleeting, dynamic, and transgressive manner. Whether we discuss the process of speciation or the symbolic meanings of nature to city folks, it is the stories of the process of their perpetual reworking that elucidates their being as part of a process of continuous transformation in which the stories themselves will subsequently take part. Latour's networks and quasi-objects need to be historicized, as following Ariadne's thread through the Gordian knot of socio-nature's networks—as Latour suggests—is not good enough if this is stripped from the *process* of its historical geographical production (Escobar 1999). Hybridization is a process of production, of becoming, of perpetual transgression. Lefebvre's insistence on temporality(ies), combined with Latour's networked reconstruction of quasi-objects, provides a glimpse of how a reworked political ecology of the city might be practised.

The production process of socio-nature embodies both material processes and the proliferating discursive and symbolic representations of nature. Therefore, if we maintain a view of dialectics as internal relations (Olman 1993; Balibar 1995; Harvey 1996) as opposed to external recursive relationships, then we must insist on the need to transcend the binary formations of nature and society and develop a new language that maintains the dialectical unity of the process of change as embodied in the thing itself. 'Things' are hybrids or quasi-objects (subjects and objects, material and discursive, natural and social) from the very beginning. By this I mean that the 'world' is a process of perpetual metabolism in which social and natural processes combine in a historical geographical production process of socio-nature, whose outcome (historical nature) embodies chemical, physical, social, economic, political, and cultural processes in highly contradictory but inseparable manners.

Figure 1.3 summarizes this argument. None of the component parts is reducible to the other, yet their constitution arises from the multiple dialectical relations that swirl out from the production process itself. Consequently, the parts are always implicated in the constitution of the 'thing' and are never outside the process of its making. In sum, then, the above perspective is a process-based episteme in which nothing is ever fixed or, at best, fixity is the transient moment that can never be captured in its entirety as the flows perpetually destroy and create, combine, and separate. This particular dialectical perspective also insists on the non-neutrality of relations in terms of both their operation and their outcome, thereby politicizing both processes and fluxes. It also sees distinct categories (nature, society, city, species, water etc.) as the outcome of materially discursive practices that are creatively destroyed in the very production of socio-nature.

It is, of course, this perspective that Harvey (1996) insists on as being the epistemological entry into the excavation of the political-ecology of capitalism. A number of analytical tools arising from this formulation are useful for the political-ecological study of water:

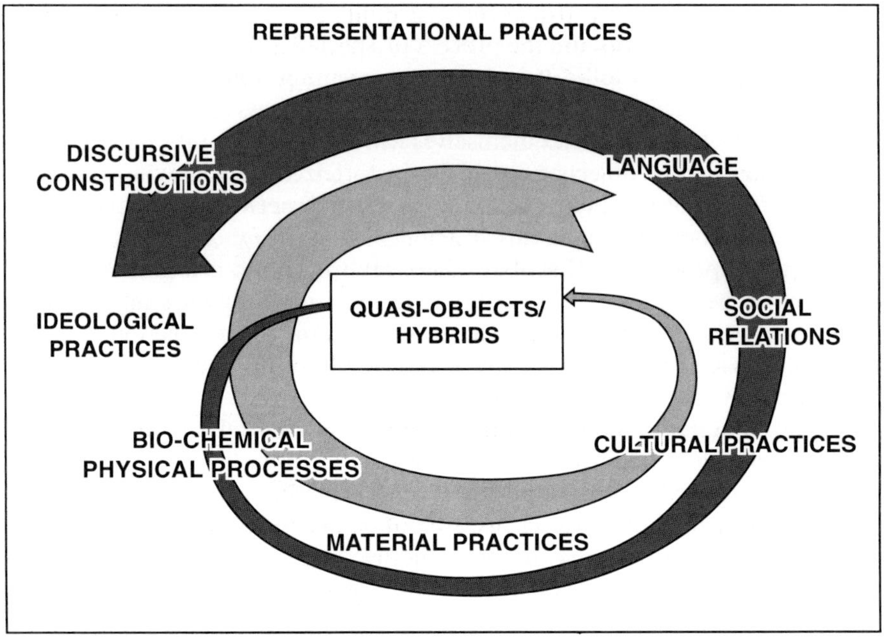

Fig. 1.3. The production of socio-nature.

1. Although we cannot escape the 'thing', transformative knowledges about water and the waterscape can only be gauged from reconstructing the processes of its production.

2. There is no 'thing-like' ontological or essential foundation (social, natural, or textual), as the process of becoming and of hybridization has ontological and epistemological priority.

3. As every quasi-object/cyborg/hybrid internalizes the multiple relations of its production, 'any-thing' can be entered as the starting point for undertaking the archaeology of her/his/its socio-natural metabolism (the production of her/his/its socio-nature).

4. This archaeology has always already begun and is never ending (cf. Althusser's infamous 'history as a process without a subject'), and is therefore always open, contested, and contestable.

5. Given the non-neutral and intensely powerful forces through which socio-nature is produced, this perspective does not necessarily lead to a relativist position. Every archaeology and its associated narratives and practices are always implicated in and consequences of this very production process. Knowledge and practice are always situated in the web of social power relations that define and produce socio-nature.

6. The notion of a socio-natural production transcends the binary distinctions between society/nature, material/ideological, and real/discursive.

The characteristics of the particular political ecological perspective on which I draw (see also Benton 1996; Keil and Graham 1998; Keil 2000; O'Connor 1998; Peet and Watts 1996; Swyngedouw 1999; Gandy 2002) can now be summarized as follows:

1. Environmental and social changes co-determine each other (Noorgaard 1994; O'Connor 1994). Processes of socio-environmental change transform both social and physical environments and produce social and physical milieus with new and distinct qualities. In other words (urban) environments are combined socio-physical constructions that are actively and historically produced, both in terms of social content and physical-environmental qualities (Escobar 1999; 2001; Latour 1993; 1999; Roberts and Emel 1992).
2. There is nothing a priori unnatural about produced environments such as cities, lakes, or irrigated fields (Harvey 1996). Produced environments are specific historical results of socio-environmental processes.
3. The type and character of physical and environmental change, and the resulting environmental conditions, are not independent from the specific historical social, cultural, political, or economic conditions and the institutions that accompany them (Swyngedouw 1997; 1999).
4. All socio-spatial processes are invariably also predicated upon the transformation or metabolism of physical, chemical, or biological components (Swyngedouw 1996*b*).
5. These metabolisms produce a series of both enabling and disabling social and environmental conditions. Indeed, these produced milieus often embody contradictory tendencies. While environmental (both social and physical) qualities may be enhanced in some places and for some people, these often lead to a deterioration of social and physical conditions and qualities elsewhere (Martinez-Allier 1991; Peet and Watts 1996; Keil 2000).
6. Processes of socio-environmental change are, therefore, never socially or ecologically neutral. This results in conditions under which particular trajectories of socio-environmental change undermine the stability or coherence of some social groups or places, while the sustainability of social groups and places elsewhere might be enhanced. In sum, the political ecological examination of the urbanization process reveals the inherently contradictory nature of the process of socio-environmental change and teases out the inevitable conflicts (or the displacements thereof) that infuse socio-environmental change.
7. Particular attention, therefore, is paid to social power relations (whether material or discursive, economic, political, and/or cultural) through which socio-environmental processes take place. It is these power geometries and the social actors who carry them out that ultimately decide who will have access to or control over, and who will be excluded from access to or control over, resources or other components of the environment. These power geometries, in turn, shape the particular social and political configurations and the environments in which we live.

8. Questions of socio-environmental sustainability thereby become fundamentally political questions. Political ecology attempts to tease out who gains from and who pays for, who benefits from and who suffers (and in what ways) from particular processes of socio-environmental change. It also seeks answers to questions about what or who needs to be sustained and how this can be maintained or achieved.

9. Political-ecological perspectives seek to unravel the nature of the social relationships that unfold between individuals and social groups and how these, in turn, are mediated by and structured through processes of ecological change. In other words, environmental transformation is not independent from class, gender, ethnic, or other power struggles.

10. It also seeks to question the actual processes of environmental reconstruction and recasting and advocates a position on sustainability that is achieved by means of a democratically controlled and organized process of socio-environmental (re-)construction. The political programme, then, of political ecology is to enhance the democratic content of socio-environmental construction by means of identifying the strategies through which a more equitable distribution of social power and a more inclusive mode of environmental production can be achieved.

While reconstructing the production processes of socio-natural networks along the lines presented above is difficult, I would maintain that such a perspective has profound implications for understanding the relationship between capitalism, modernity, ecology, space, and contemporary social life. It also has implications for transformative and emancipatory ecological politics.

1.5 Emancipatory hybrids

A number of recent contributions to this debate have begun to address this problematic in one way or another. William Cronon (1991), for example, in *Nature's Metropolis*, tells the story of Chicago from the vantage point of the socio-natural processes that transformed both city and countryside and produced the particular political ecology that shaped the transformation of the mid-West and produced a particular American socio-nature. While symptomatically silent about the myriad of struggles that have infused this process (African-American, women's, and workers' organizations and struggles are notoriously absent from or marginalized in his narrative), the book marks interesting and powerful pointers on the way to a political ecology of the city.

Mike Davis (1990), for his part, in *City of Quartz* and other recent publications (Davis 1995; 1998) suggests how nature and society become materially and discursively constructed in and through the dialectics of Los Angeles' urbanization process, and how multiple social struggles have infused and shaped this process in deeply uneven, exclusive, and empowering/disempowering ways. For him, homelessness and racism combine with pollution, earthquakes and water scarcity as the most acute socio-ecological problems that have been produced through the particular form of post-industrial capitalist development that has shaped Los Angeles as the Third World Megalopolis. Indeed, the history of Los Angeles' urbanization process indicates how the socio-ecological transformation of desert lands, the manufacture of an orchard socio-nature, and subsequent construction of 'silicon' landscapes is paralleled by urbanizing, capturing and controlling ever larger and distant catchments, by speculatively pushing the frontier of 'developable' land further outwards and by an ever-changing, but immensely contested and socially significant (in terms of access and exclusion; empowerment/disempowerment) choreography of national laws, rules, and engineering projects (Worster 1985; Gottlieb and Fitzsimmons 1991). Of course, as the deserts bloomed, ecological and social disaster hit: water scarcity, pollution, congestion, and lack of sewage-disposal combined with mounting economic and racial tensions and a rising environmentalism (O'Connor 1998: 118; Gottlieb 1993; Keil and Desfor 1996; Keil 1998). The rhetoric of disaster, risk, and scarcity often provided the discursive vehicles through which power-brokers could continuously reinvent their boosterist dream. Picturing a simulacrum of drought, scarcity and a return to the desert produced a spectacularized vision of the dystopian city whose fate is directly related to faith in the administrators, engineers, and technicians who make sure the tap keeps flowing and land keeps being 'developed'. The hidden stories of pending socio-ecological disaster provide the ferment in which local, regional, and national socio-natures are combined with engineering narratives, land speculation, and global flows of water, wine, and money.

Matthew Gandy (2002) narrates with great skill and in exquisite detail the reworking of nature in New York City, a reworking that is simultaneously material and physical, and embedded in political, social, and cultural framings of nature. At the same time, the myriad power relations and political strategies that infuse the socio-environmental metabolism of New York's socio-nature are meticulously excavated and taken to centre stage in the reconstruction of contemporary New York as a cyborg city.

In the next chapter, we shall tentatively explore this perspective somewhat further. Our vantage point will be the circulation of water, its insertion in the metabolism of the city and in the political ecology of the urbanization process. The flow of water, in its material, symbolic, political and discursive constructions, embodies and expresses exactly how the 'production of nature' is both arena and outcome of the tumultuous reordering of socio-nature in ever-changing and intricate manners. This flow of water as a socio-environmental metabolic process and its historical geographical production process will be used as an entry point to excavate the process of modern urbanization in Guayaquil, Ecuador. The production of the city as a cyborg, excavated through the analysis of the circulation of hybridized water, opens up a new arena for thinking and acting on the city; an arena that is neither local nor global, but that weaves a network that is always simultaneously deeply localized and extends its reach over a certain scale, a certain spatial surface. The tensions, conflicts and forces that flow with water through the body, the city, the region and the globe show the cracks in the lines, the meshes in the net, the spaces and plateaus of resistance and of power.

2

The City in a Glass of Water: Circulating Water, Circulating Power

> In the period of which we speak [eighteenth-century France], there reigned in the cities a stench barely conceivable to us modern men and women. The streets stank of manure, the courtyards of urine, the stairwells stank of mouldering wood and rat droppings, the kitchens of spoiled cabbage and mutton fat; the unaired parlours stank of stale dust, the bedrooms of greasy sheets, damp featherbeds, and the pungently sweet aroma of chamber-pots. The stench of sulphur rose from the chimneys, the stench of caustic lyes from the tanneries and from the slaughterhouses came the stench of congealed blood. People stank of sweat and unwashed clothes; from their mouths came the stench of rotting teeth, from their bellies that of onions, and from their bodies, if they were no longer very young, came the stench of rancid cheese and sour milk and tumorous disease. The rivers stank, the marketplaces stank, the churches stank, it stank beneath the bridges and in the palaces. The peasant stank as did the priest, the apprentice as did his master's wife, the whole of the aristocracy stank, even the King himself stank, stank like a rank lion, and the Queen like an old goat, summer and winter.
>
> Patrick Suskind 1987: 3

2.1 The socio-hydrological metabolism of urban water

In recent years, an impressive body of work has emerged in the wake of the resurgence of the environmental question on the political agenda, addressing the environmental implications of urban change or issues related to urban sustainability (Haughton and Hunter 1994; Satterthwaite 1999). In many, if not all, of these cases, the environment is defined in terms of a set of ecological criteria pertaining to the physical milieu. Both urban sustainability and the environmental impacts of the urban process are primarily understood in terms of physical environmental conditions and characteristics.

We start from a different position. As explored in Chapter 1, urban water circulation and the urban hydrosocial cycle are the vantage points from which the urbanization process will be analysed in this book. In this Chapter, a glass of water will be my symbolic and material entry point into an—admittedly somewhat sketchy—attempt to excavate the political ecology of the urbanization process.[1] If I were to capture some urban water in a glass, retrace the networks that brought it there and follow Ariadne's thread through the water, 'I would pass with continuity from the local to the global, from the human to the non-human' (Latour 1993: 121). These flows would narrate many interrelated tales: of social and political actors and the powerful socio-ecological processes that produce urban and regional spaces; of participation and exclusion; of rats and bankers; of water-borne disease and speculation in water industry related futures and options; of chemical, physical, and biological reactions and transformations; of the global hydrological cycle and global warming; of uneven geographical development; of the political lobbying and investment strategies of dam builders; of urban land developers; of the knowledge of engineers; of the passage from river to urban reservoir. In sum, my glass of water embodies multiple tales of the 'city as a hybrid'. The rhizome of underground and surface water flows, of streams, pipes and networks is a powerful metaphor for processes that are both social and ecological (Kaïka and Swyngedouw 2000). Water is a 'hybrid' thing that captures and embodies processes that are simultaneously material, discursive, and symbolic.

Water is of course biochemically vital, embodies deep social meaning and cultural value, and internalizes powerful socio-economic and physical relations. It is increasingly becoming part of a new 'accumulation strategy' (Fitzsimmons 1989; Katz 1998) through the privatization of waters that were often part of a common or collective good. Yet life in any form is scarcely imaginable in the absence of water. The socio-natural production of the city is predicated upon some system of circulating water. The multiple temporalities and interpenetrating circulations of water (the hydrological cycle, canalization and distribution networks, treatment stations, etc.) illustrate the perpetual metabolism and mobilizations of water. Moreover, there are water problems of gigantic dimensions worldwide (Ward 1997; Petrella 1998), with over one billion people lacking access to reasonably potable water. Mega-cities in the developing world suffer from immense water shortages, while the water metabolism in developed cities threatens the very metabolism of urban life as pollutants of all kinds challenge the very sustainability of the capitalist city and the metabolism of social and biological life (Gleick 1993; Postel 1992). In addition, water carries powerful symbolic meanings (health, purity, naturalness), which in recent years have been successfully 'mined' by a burgeoning global multi-billion dollar mineral water industry.

Our glass of water relates all things/subjects in a network, connects the most intimate of socio-spatial relations, inserts them into a mesmerizing political economy of urban, national and international development, and is part of a

[1] For fuller accounts of aspects of this argument, see Swyngedouw (1995*a*, *b*; 1996*b*). For a more detailed reconstruction of various interwoven water narratives, see Swyngedouw (1999).

chain of local, regional, national, and global circulations of water, money, texts, and bodies. In this sense, I would insist that we can reconstruct, and hence theorize the urbanization process as a political ecological process with water as the entry point: water that embodies biochemical and physical properties, cultural and symbolic meanings, and socio-economic characteristics simultaneously and inseparably. These multiple metabolisms of water are structured and organized through socio-natural power relations—relations of domination and subordination, of access and exclusion, of emancipation and repression—which then become etched into the flow and metabolisms of circulating water. This flow of water produces not only a physical geography and a material landscape, but also a symbolic and cultural landscape of power. The waterscape is a liminal landscape (to use Zukin's words (1991)) in which the cyborg character of the transgression between socio-nature and nature's society is perpetually emptied out, filled in again, and transformed (see Keil 1994). This circulation of water is embedded in and interiorizes a series of multiple power relations along ethnic, gender, and class lines (see Swyngedouw 1995*a*). These situated power relations, in turn, swirl out and operate at a variety of interrelated geographical scale levels, form the scale of the body upward to the political ecology of the city to the global scale of uneven development. The capturing, sanitizing, and biochemical metabolizing of water to produce 'urban' drinking water simultaneously homogenizes, standardizes, and transforms it into a commodity as well as into the real/abstract homogenized qualities of money power in its manifold symbolic, cultural, social, and economic meanings. The struggle for water, and competition over access to it, capture wider processes of the political ecology of urbanization, and produce the multiple and scaled power relationships to which we will now turn our attention.

2.2 Urbanizing water

It is, of course, fairly trivial to say that the urbanization process is predicated upon myriad socio-ecological transformations that affect the geographies of places both nearby and far away (Cronon 1991; Hundley 1992; Gottlieb and Fitzsimmons 1991). This intense socio-environmental transformation is required to 'sustain' the dynamics of contemporary urban change, resulting in the formation of various new environments—from concrete urban landscapes to aquatic ecosystems around reservoirs. The process of urbanization is both a historically specific accumulation of socio-environmental processes and the arena through which these transformations take place. The material and imagined bond between water and urban space as social creations provides an ever-changing material and metaphorical surface that produces a narrative of the history and geography of water. Water has always possessed powerful connotations and conveyed important symbolic messages. 'Naturalness', virginity, healing, and purification have often been associated with water, while water spectacles have in many ways testified to the power and the glory of various kinds of (urban) elites (Moore and Lidz 1994; Schama 1995). For example, the cultural links between female nudity and the tap water of the bathroom began to be formed in the second half of the nineteenth century, as the sprinkling of water from an intricately engineered network of pipes over the naked (female) body became part of the fantasy of sexual intimacy (Illich 1986: 1). Simultaneously, water became a commodity, expressing the social relationships within the space through which it circulated and to which it gave form and content. The biological necessity of water ensured that urbanization was predicated upon organizing, controlling, and mastering its socio-natural circulation. For example, in Mexico City, 60% of all urban potable water is distributed to 3% of the households, whereas 50% make do on 5%. In Guayaquil, 65% of the urban dwellers receive 3% of the produced potable water at a price that is at least two hundred times higher (20,000%) than that paid by the low-volume consumer connected to the piped urban water network. The mechanisms of exclusion from and access to water manifest the power relationships through which the geography of cities is shaped and transformed. The history of the urbanization of water illustrates the intricate ways in which the image and reality of water access and use is bound up with social transformations and the formation of the modern city. The urbanization of life and the urbanization of water are intimately connected. But this urbanization of water as we know and accept it today is dependent on viewing water circulation as a perpetual movement from the 'natural' source to and through the city via a series of social and physical metabolic transformations, until it ends up once again at the source, reintegrated into the flow of 'natural' water.

Cities first became dependent on water flowing through aqueducts that pierced the city wall, or from wells penetrating the earth. Nine major aqueducts, with a total length of over 400 kilometres, supplied approximately 400 litres of water per capita per day to ancient Rome, which had a population of approximately a million by AD 100. However, one fifth of this water was assigned to the emperor, whilst another two-fifths fed the city's 591 fountains and dozen public baths (see Scobie 1986). In contrast, in 1823, London, Frankfurt, and Paris had 3 litres per capita per day, a figure which had only risen to approximately 40 litres by 1936 (Mumford 1961); a volume less than that found in many cities in the colonial or post-colonial world at that time. Water brought from afar to ancient non-Roman cities was usually absorbed by city soil, as sewers that channelled piped water remained the exception. In Rome, water from fountains flowed over the streets and into the Tiber.

2.2.1 The invention of circulation

The concept of 'water circulation', with water following a given path into, through, and finally out of the city by the sewers remained foreign to western urban imaginations, spatial representations, and engineering systems until the nineteenth century. Modern urbanization, highly dependent on the mastering of circulating water, is linked with the representation of water as a circulatory system. Before the 'discovery' of circulatory systems, the movement of water was seen merely as evaporation: the separation of the 'spirit' from the 'water' (Goubert 1989). Phlogiston theory, the representation of the respiratory system, plant growth, and the physiocratic view of the production of material wealth all indicate that Renaissance people did not conceive of 'circulation' as an infinite cyclical process. The idea of a material flowing forever back to its own source signalled a major breakthrough. When William Harvey stated his ideas of the double circulation of blood in the vascular system of the body in 1628, a revolutionary insight came into being which would begin to permeate and dominate, both metaphorically and materially, everyday life, engineering and academic practice for centuries to come.[2] By the end of the seventeenth century, medical practice had accepted the idea of the circulatory (metabolic) system, and by the nineteenth century the metabolic circulation of chemical substances and organic matter became increasingly accepted and began to form the basis of modern ecology.

At the beginning of the eighteenth century, the term 'circulation' of liquids had become established in many sciences, from the flow of sap in plants to the circulation of matter in chemical reactions (Teich 1982). From about 1750, wealth and money were spoken of as 'circulating' as though they were liquids, flowing incessantly as part of a process of accumulation and growth, and society began to be imagined as a system of conduits (Sennett 1994). 'Liquidity' arose as a dominant metaphor after the French Revolution: ideas, newspapers, gossip and—after 1880—traffic, air, and power were said to 'circulate'. Montesquieu in *Lettres Persanes* (p. 117) speaks of '[T]he more "circulation" the more wealth' and in *l'Esprit des Lois* of '[M]ultiply wealth by increasing "circulation" '. Jean Jacques Rousseau (1766) refers to '[T]his useful and fecund circulation that enlivens all society's labour' and to 'a "circulation of labour" as one speaks of the circulation of the money' (cited in Illich 1986: 23). Adam Smith and, in particular, Karl Marx conceived of a capitalist economy as a metabolic system of circulating money and commodities, carried by and structured through social interactions and relations, in which accumulation is dependent on the swiftness by which money circulates through society. Each hiccup or deceleration of circulation threatened to unleash the infernal forces of devaluation, crisis, and chaos. Society's wealth and the relationships of power on which wealth is constructed were seen as being intrinsically bound up

[2] The first person to suggest the circulation of blood in the arterial system was apparently Ibn-an-Nafiz (physician, born in Baghdad and died in Cairo in 1288) (Illich 1986: 40). The idea of circulation remained alien to the imagination of sixteenth-century Europeans. Two sixteenth-century scientists suspected what Harvey would later discover: Servetus (a Spanish genius and heretic burnt by Calvin—he also edited Ptolemy's geography in Lyon and was a student of Vesalius in Paris) and Realdus Colombus of Padua (also a student of Vesalius). Harvey was a student of Vesalius in 1603.

with and expressed by the 'circulation speed' of money in all its forms (capital, labour, commodities) (Harvey 2002). The development and consolidation of circulating money as the basis for material life and the relations of domination and exclusion through which the circulation of money is organized and maintained shapes the 'urbanization of capital' (Harvey 1985) and, inevitably, the 'urbanization of water'.

The status of water within city space changed as the purifying and cleaning power of water began to dominate the metaphorically constructed healing and rejuvenating water rituals of baptism and exorcising miasma. By the mid-nineteenth century some British architects begin to speak of the inner city using the same metaphor of circulation, and in 1842 Sir Edwin Chadwick formulated the ideology of circulating waters effectively for the first time. Chadwick (1842) presented a report on the sanitary conditions of the labouring population of Great Britain which Lewis Mumford has called 'the classic summary of paleo-technic horrors'. In his report, Chadwick imagined the new city as 'a social body through which water must incessantly circulate, leaving it again as dirty sewage'. Water ought to 'circulate' through the city without interruption to wash it of sweat, excrement, and waste. The brisker this flow, the fewer stagnant pockets that breed pestilence there are and the healthier the city will be. Unless water constantly circulates through the city, pumped in and channelled out, the interior space imagined by Chadwick can only stagnate and rot. This representation of urban space as constructed in and through perpetually circulating flows of water is conspicuously similar to imagining the city as a vast reservoir of perpetually circulating money. In fact, Chadwick's papers were published under the title *The Health of Nations* during the centenary commemoration for Adam Smith (Chadwick 1887). Like the individual body and bourgeois society, the city was now also described as a network of pipes and conduits. The brisker the flow, the greater the wealth, the health, and hygiene of the city would be (Vigarello 1988). Just as William Harvey redefined the body postulating the circulation of the blood, so Chadwick redefined the city by 'discovering' its needs to be constantly washed (Illich 1986: 45). And of course, Baron von Haussman, the engineer who masterminded the reorganization of Paris's cityscape also successfully mobilized the metaphor of 'circulation' to impress and convince the city's leaders of the necessity of his grandiose project (Gandy 1999); a project that would permit all sorts of flows, from sewage to people and commodities, to move more swiftly through the city. Later, David Harvey (1985) would analyse the circulation of capital and its urbanization as a perpetuum mobile channelled through a myriad of ever-changing production, communication, and consumption networks, driven by a motley crew of financial speculators, profit-seeking capitalists, visionary urbanists, and enlightened elites striving to modernize and 'civilize' urban life.

With this enlightenment idea of 'circulation', the utopia of the odourless, clean, purified city appears:

> [The] effort to deodorize utopian city space should be seen as one aspect of the architectural effort to 'clear' city space for the reconstruction of a modern capital. It can be interpreted as the repression of smelly persons who unite their separate auras to create a smelly crowd of commonfolk. Their common aura must be dissolved to make space for a new city through which clearly delineated individuals can circulate with unlimited freedom. For the nose of a city without aura is literally a 'Nowhere', a u-topia'. (Illich 1986: 53)

The image and practice of water, now disciplined and harnessed in circulatory urban water systems, was profoundly transformed. Defecating became a sex-specific activity for the first time in history towards the middle of the eighteenth century, as separate latrines for men and women were set up—but only for special occasions (Corbin 1994). At the end of that century, Marie Antoinette had a door installed to her lavatory, thus turning the act of defecation into an intimate function (Illich 1986: 57). The degree to which it is practised in private also signals a certain social status and an embracing of superior civic morality (Vigarello 1988). On 15 November 1793, the French revolutionary convention solemnly declared each man's right to his own bed, thus enshrining the right to be surrounded by a buffer zone protecting the citizen from the aura of others. The private bed, stool, and grave became requisites of a citizen's dignity. Children began to learn that hygiene and sanitary activities are a solemn, private process (Goubert 1989), again indicating a profound redefinition of the self and the body in the 'utopian' urban space.

The toilette of the whole city was undertaken in parallel with the privatization of body relief and the attempt to retrench people's auras, reducing each other to an odourless point in the new civic space. This culminated in the modern design principles, heralding clean air, ventilation, pure water, and treated sewage (Kaïka and Swyngedouw 2000). Water became a detergent of smell, as one could move up the social ladder only through eliminating body smells. It was not until the nineteenth century that soap became associated with body laundry and the social repression of smell became an element in the class struggle of the elite in search of 'cultural capital' to distinguish themselves from the 'smelly' commoners. Shortly afterwards, perfume and the 'domestication of aura' (Illich 1986: 62) became employed in the act of seduction, no longer merely covering body smell, but artificially providing secondary sexual characteristics to the new 'human' body. Like so many other characteristics—including work, health, and education—smell, too, is henceforth conceived as an abstract quality that is 'naturally' polarized into a female and a male type: she smells of violets and roses and he of leather and tobacco. The old 'toilette' of the eighteenth century referred to a hydrophobic process (Illich 1986: 65) of combing, grooming, powdering, applying make-up and perfumed cosmetics, dressing, and finally receiving visitors in the boudoir. Frequent cleansing by means of water was not part of the toilette until the nineteenth century, but by the 1830s the word had come to mean the sponging of a naked body, invariably represented as belonging to a woman. From decade to decade, the amount of water used in the procedure increased. The toilette came to mean a tub bath, and around 1880, the industrial production of enamel paints replaced expensive copper with iron or zinc vessels and brought the tub within reach of simple families (Wright 1960). 'Toilette' retired behind closed doors (together with perfuming, shitting, and shaving), and began to involve the flow of tap water to carry soapsuds and excrements to the sewer (Goubert 1989). When the first urban water system in Guayaquil was installed, for example, the urban elites brought finely decorated lavatories and washing bowls from their trips to Europe to testify to their newly acquired sanitized civic conditions. Lower classes and indigenous people visited the houses to marvel at these symbols of a new elite urban order. The total bathroom was not installed overnight. It is revealing that the place in which the modern body is integrated into the circulation of city waters is called the 'bath'-room. Its initial use, according to the Oxford English Dictionary (first mentioned only in the supplement of 1972) is traced to 1888. The choice of this term indicates that the identification of nature and the nude, which Ingres, Courbet, Degas, and Renoir had painted as taking place in rivers, under waterfalls, or in an 'oriental' hammam, was actually performed in the intimacy of the toilette (Illich 1986: 66). Public space became increasingly hydrophobic and the public body in the western city desperately tried to cover itself as protection from public waters and public gazes alike. Indeed, as Vigarello (1988: 216) attests, 'the exclusion of others became an obligatory element in the cleanliness of the elite'.

In sum, the increasingly commodified domestication of water announced the withdrawal of the urban elite body and bodily hygiene from the public or semi-public sphere and its retreat into the privacy and intimacy of the bathroom and the toilet. The hydrophobic public spaces were replaced by hydrophilic private spaces as bodily encounters were relegated to specially designed places. This, in turn, redefined the body and bodily relations. Nudity and exposing the naked body to the 'elements' became improper and uncivilized. The new sanitized and deodorized (washed) urban body in a sanitized urban public civic space redefined both class and gender relations. Images of (predominantly female) sexuality began to revolve around the secrets, intimacy, and eroticism associated with the bathroom, the toilet, and the sprinkling of domesticated water over the naked body (Corbin 1994). Of course, the newly de-odorized urban body, embodying quite literally a new civic, modern ideal, carried by an urban bourgeoisie that was becoming quickly self-confident of its new role, became re-odorized in new ways, expressing cultural distinction and power differentiation (Bourdieu 1986). But this new urban civic body also separated the sanitized bodies of the new urban elites from the peasant reeking of manure and the sweaty proletarian. Class and gender relations became impregnated with smell and odour and the body aura became an element in cultural and social differentiation and power relations (Suskind 1987; Corbin 1994; Rindisbacher 1992).

Urban waterworks signalled this new class and gender differentiation. The mechanisms of exclusion from and access to *unlimited* quantities of potable water were cemented into the water engineering system itself and remain like this until this very day. In many Third World cities, for example, the elites, clustering around the water reservoirs, had and have unlimited access to water, which in addition to the above cultural distinctions, turned this into significantly longer life expectancy and into valued symbols of cultural capital and social power. In Guayaquil, permanently irrigated tropical gardens separate their often militarized urban oases in their gated communities from the urban desert that surrounds them, fountains in the courtyards testify to their social position. Images of the smelly peasant and unhygienic indigenous population re-enforce the position of water as an integral element of social power in the city and part of the process of the urbanization of nature. Nevertheless, water-related illnesses and deaths remain the major cause of infant mortality for most of the world's population. In short, the urban ecological conquest of water and the fusion of water circulation with the urbanization process (for a vivid account see J. Vernes's futuristic account of Paris (1994)), its commodified domestication and related processes of access to and exclusion from access brought water squarely into the realm of urban social power.

2.2.2 Social power and water control

The domestication of water and the privatization of bodily hygiene were predicated upon and paralleled by an increasing commodification of water. The urbanization of water necessitated both ecological transformation (capturing water from underground aquifers or distant catchments, engineering its flow, negotiating geopolitical relations, transforming its chemical and biological properties and so forth) *and* social transformation. Indeed, the very homogenization and standardization of 'potable' urban water propelled the diverse physical, chemical, and biological 'natural' flows and characteristics of nature's water into the realm of commodity and money circulation with its abstract qualities and concrete social power relations. 'Potable' water became legally defined and standardized. Biochemical and physical treatment (adding or extracting substances) was required to homogenize water according to 'scientific' politically and socio-culturally defined norms that were enshrined in binding legislation. Homogenization, standardization, and legal codification are essential to the commodification process.

The urban conquest and commodification of water brought H_2O squarely into the sphere of money and cultural capital and its associated power relations, and redrew socio-natural power relations in important new ways. The use of water for the cleaning of the body and the use of water for the 'toilette' of city spaces go hand-in-hand but not at the same pace in all modern nations. However, the urbanization of water through vast engineering systems of production, conduction, and distribution became an inherent element underpinning the urbanization of society in the nineteenth and twentieth centuries. The modern city had become a rhizome of networks and conduits.

The 'modern' engineering systems through which water is mastered and becomes commodified demand large capital investments with installations that have a long life-span (sometimes 50 to 100 years) and an immense infrastructure system that guides the circulation of water in an interconnected way over a large scale, often covering entire regions (Montano and Coing 1985). It is clear that such a system requires some form of central control and a coordinated, combined but detailed division of labour (see Wittfogel 1957; Worster 1985). In addition, the quantity, quality, and regularity of the circulating waters are determined by the weakest link in this detailed technical and social division of labour. Sufficiently large amounts of capital have to be amassed and sunk into the construction of massive engineering systems with long turnover times. The early private capital-based urban 'watering' initiatives were gradually replaced by primarily state-funded investments in public waterworks, managed by large public or mixed public-private companies (Lorrain 1995) (see below). Circulating capital had to be captured and organized in fixed physical infrastructures that would permit the 'free' circulation of clean water (as well as of the waste waters).

In addition, the processes of water production, conduction, and distribution are necessarily spatially structured, shaping and being shaped by urban and regional geographies. Producing and providing water is essentially and necessarily a deeply localized activity, while transporting water is a difficult—and costly—process. This double tendency of modern water systems towards centralization and central control on the one hand and the necessarily localized character of all parts of its circulation process on the other, works itself out in very contradictory and conflicting ways as will be documented in our study of Guayaquil. Although geo-climatic conditions such as the availability and type of natural water resources and pluvial regimes, as well as settlement patterns, are of a great importance for the organization of water management systems, these physical characteristics cannot be separated from the organization of human relations. Indeed, the relative scarcity of usable water will only influence the mode of water management to the extent that social groups will enter into competition for its utilization and that relations of cooperation and relations of power will translate themselves into specific institutional, managerial, and technological systems (Anton 1993). Montano and Coing (1985: 8) summarize this succinctly:

> The management of water is, therefore, always the result of the social relationships which crystallise around its appropriation and its usage. It varies in function of both the geo-climatic constraints and the relationships of power between users.

The social struggle around water is evidently the result of the deeply exclusive and marginalizing political, economic, and ecological processes that drove the expansion of the city. The urbanization process is predicated upon the mastering and engineering of nature's water, with the ecological conquest of water as a necessary prerequisite for the expansion and growth of the city. At the same time, the capital required to build and expand the urban landscape is also generated through the political-ecological transformation of the city's hinterland (Swyngedouw 1996*b*; 1997). In short, the political ecological history of many cities can be written from the perspective of the need to urbanize and domesticate nature's water and the parallel necessity to push the ecological frontier outward as the city expanded (see Chapters 4 and 5). As such the political ecological process produces both a new urban and rural socio-nature. The city's growth, and the process of water urbanization is closely associated with successive waves of ecological conquest and the extension of the urban socio-ecological frontier. Local, regional, and national socio-natures are combined with engineering narratives, economic discourses and practices, land speculation, the geopolitics of water, and global money flows. Investments in bottled water companies, speculation in water industry-related financial instruments and global/local hydrological cycles fuse together in the production of hybridized waters and cyborg cities. Water circulation and the urbanization of water thus become deeply entrenched in the political-ecology of the local and national state, the international divisions of labour and power, and the local regional and global hydrological climatic cycles.

In short, the urbanization of water and the social, economic, and cultural processes associated with the domestication of water brought access to nature's water squarely into the realm of class, gender, and cultural differentiation and made water subject to an intense struggle for control and/or access. The commodification of water, in turn, placed the circulation of water directly in the sphere of money circulation, which consequently made access to water dependent on positions of social power, both economically and in terms of gender and culture. Although the particular geographical and institutional configurations vary significantly from city to city and from country to country depending on the particular combination of physical and institutional factors, the twentieth-century urbanization process and the accompanying expansion of water use significantly affected the spatial choreography of urban water circulation (Graham and Marvin 2001). For each expanding city, the physical territorial basis on which the successful watering of the city rests needs to expand as the city grows, in quantitative as well as in qualitative terms (Hundley 1992). Either new untapped water reserves have to be incorporated in the urban water cycle or existing water supplies tapped more intensely. In the case of aquifer water, this leads either to a problem of generalized over-pumping which outstrips the natural recharge capacities of aquifers or to a gradual decline in the quality of aquifer waters. The geographical expansion of the ecological footprint of urban water not only transforms places and environments far removed from the city, but also intensifies conflicts with other users over limited water supplies. From the vantage point of the early twenty-first century, there is increasing evidence that the sustainability of urban development was bought at the expense of an expanding water frontier and of geographically widening the sphere of impact of the urban water cycle, leading to socially conflicting and socio-ecologically unsustainable practices of expanding resource extraction and intensified struggle for control or access. In what follows, some of the above arguments will be marshalled to elucidate the central trends that characterize contemporary urban water management systems. The key points of tension, conflict, rupture, and/or potential crisis will be discussed.

2.3 Critical moments in the contemporary urban hydrosocial cycle

2.3.1 The shifting political economy of water

Despite the recent debates that have raged over the privatization of urban water systems—debates that are often couched in terms of an inevitable and necessary adaptation of national policies to the requirements of a global and deregulated neo-liberal world economic order—shifting configurations of public–private partnerships have characterized changes in the urban water sector since the inception of modern supply systems (Hassan 1998; Castro and Swyngedouw 2000). The boundaries between public and private spheres have changed from time to time and relative balances have moved more to one side or the other, but in common with many other public goods, the water sector has been customarily characterized by a certain articulation of public and private actors.

Most international studies show that the organization of urban water supply systems can be broadly divided into four stages (Hassan 1998). The first of these was the period up to the second half of the nineteenth century, when most urban water supply systems consisted mainly of relatively small private companies providing parts of the city (usually the richer parts) with water of varying quality. Water provision was socially and spatially highly stratified and water businesses were aimed at generating profits for the investors. In colonial cities, waterworks tended to concentrate on the areas where the colonial elites, both domestic and imperial, lived and worked, although a variety of other local water distribution mechanisms were in place, including wells, fountains, and commercial door-to-door selling.

This was followed by a period of municipalization, primarily prompted by concerns over deteriorating environmental conditions and calls for a sanitized city. In Europe, this took the form of municipal socialism with its concern for providing essential public goods at a basic, often highly subsidized, rate (Laski, Jennings, and Robson 1935; Millward 1991). Profitability was without any doubt a secondary concern and subsidies came from the general tax income (from either the local and/or the national state). This municipalization was also supported by local elites who realized that their own health and environmental conditions were negatively affected by deteriorating environmental standards in the city. Water supply systems were consolidated, leading to a citywide standardized coverage of domestic water supply, coupled with attempts to produce comprehensive sewage-disposal systems (albeit without treatment of sewage waters). In this period, the large cities in the developing world developed their water systems at a rate that was comparable with and occasionally even faster than those for comparably sized cities in the developed world.

The third phase started approximately after the First World War, and turned the water industry, together with other major utility sectors (such as electricity and telecommunications), into a growing national concern. The national state was to take a much greater responsibility in providing public services, although with a varying degree of intensity of control, regulation, and capital input. Water infrastructure became—together with other major infrastructure works and programmes—part of a Fordist–Keynesian state led social and economic policy. The investments in grand infrastructure works (dams, canals, technological networks) were part of an effort to generate and/or support economic growth, while simultaneously assuring a relative social peace by means of redistributive policies (Amin 1994; Moulaert and Swyngedouw 1987; Gandy 1997). Three objectives were central to this Fordist period of water expansion: the creation of jobs; the generation of demand for investment goods from the private sector; and the provision of basic collective production and consumption goods (like water, education, housing) at a subsidized price for wage workers and industry alike. The combined result produced the classic model of state involvement in the post-war Keynesian expansion strategies. These trends can be identified around the world, although a widening gap began to manifest itself between Western cities and cities in the rest of the world. In particular, as will be documented in Chapter 6, waterworks for cities in the developing world became structurally dependent on global investment flows, notably through bilateral or multilateral loans. This dependency would ultimately, from the late 1970s onwards, wreak havoc when the debt crisis erupted and state-based investments were significantly curtailed (Montúfar 1990). During this post-war 'Fordist' era, the state played an ever-increasing role, nationalizing water provision in some instances, financing infrastructure projects, and generally increasing regulatory intervention, although in some cases management remained under the auspices of sub-national or municipal authorities. It was indeed also during this period that a variety of regulatory bodies (for social, economic, quality, or environmental regulation) were established, usually by and at the level of the national state.

The fourth and most recent phase is associated with the demise of state-led economic growth and the subsequent transition to post-Fordist or flexible forms of economic development and governance. This started approximately with the global recession of the 1970s, and represents a time of radical changes in public/private interplay in the water sector (Estache, Gomez-Lobo, and

Leipziger 2001; Bakker 2003). First of all, mounting economic problems—in the context of high public social and investment spending—resulted in growing budgetary difficulties for the national (and often also local) state. This necessitated a serious consideration of the direction of state spending, and resulted in a reduction of expenditures in the welfare sector and in supporting industrial sectors or infrastructure that ran structural deficits. Low prices, subsidized investments in water supply systems, and ageing water infrastructure put greater pressure on state budgets, a situation compounded by a continued increase in demand for water. Secondly, the call for greater competitiveness as a means to redress the economic crisis of the 1970s and early 1980s prompted a quest for efficiency gains and greater productivity through cutting bureaucracy, deregulating labour markets, and increasing investment flexibility (Bakker 2002). This, in turn, was accompanied by privatization tendencies as a means to achieve both of the above solutions to the crisis of Fordism. Moreover, the growing globalization of the economy and the accompanying changes in the nature of competition, the greater availability of private capital achieved by means of deregulation and de-territorialization of financial markets, and the imposition of strict budget norms (by the European Union, World Bank, or IMF) further shifted the balance in favour of the private sector. Thirdly, the standard democratic channels of government, which were often infused with the active lobbying power of social organizations (particularly unions), proved to be a considerable barrier to implementing swift policy changes. The political-economic configuration has, consequently, changed in important ways, resulting in new institutional arrangements (see below) permitting a more business- or market-oriented management more in tune with profit-making strategies (Swyngedouw, Page, and Kaïka 2002). Fourthly, capital began to search for new frontiers to incorporate. Nature in all its forms (including the production of new genetic materials) became part and parcel of new accumulation strategies. Water presented itself as a possible new frontier to cross, a potential source for turning H_2O into money and profit. Private accumulation through 'dispossession', the privatization of what until recently had been primarily common or collective goods, became a favoured corporate strategy to seek out new investment niches (Bakker 2002). Finally, growing environmental problems and, consequently, the increasing number of actual and potential conflicts in the management and regulation of the water cycle proved to be a serious challenge for traditional forms of organization and implementation of water-related activities. Particularly in a context of more vocal and powerful civil society-based environmental groups, systems of governance needed to become more sensitive to these issues. Questions of restricting or controlling demand (demand management) as a strategy for lower water consumption and hence taking away the pressures on expanding the urban water resource base became more loudly heard. The internalization of all these tensions within what remained a fundamentally state-owned and state-controlled sector became increasingly difficult (Swyngedouw 1998).

The combined effect of the above processes and dynamics resulted in a shift from water sectors that were state-managed to ones which were more in tune with globalized market forces and the imperatives of a competitive privatized economy (Kazemir, Leinin, and Schaub 2002). This represents both a practical as well as an ideological and discursive shift, and occurred with varying degrees of intensity in different countries. In some cases, actual privatization of water production and delivery took place (e.g. Buenos Aires, Jakarta, or the failed attempts in Cochabamba (see Crespo 2002*a*)), whereas in other cases (e.g. Amsterdam) publicly owned companies were increasingly required to act strategically, managerially, operationally, and organizationally as private companies with an eye towards potential future privatization. In addition, water businesses are now often part of global multi-location companies, or part of larger multi-utility companies such as Vivendi, RWE, or the late Enron.

The debate over privatization, and the privatization process itself, have had and will continue to have profound implications in and for the water sector and beyond.[3] The recent shift towards turning H_2O into a commodity has dramatically changed the social and political meaning and cultural valuation of water. First of all, water is turned into profits and capital accumulation by private or public/private institutions. Supplying water therefore becomes a means to achieve an economic end: economic growth and profit maximization. To the extent that private companies do this, water-related activities become solely a strategic element within companies that are rapidly becoming multi-utility and international. Second, non-economic uses and functions of water then have to be regulated by governmental institutions that often face serious opposition, conflict or other constraints in the face of powerful private agencies. Moreover, it becomes increasingly difficult, if not impossible, to integrate water policies within a wider urban, social, or economic policy involving cross-subsidization, alternative uses of water, or a socially stratified policy. Third, this shift inevitably entails a change in the geometry of social power. Private actors and companies become much more powerful voices in strategic water-related decisions, at the expense of other civil society organizations or of the state. Fourth, while the water cycle operates along temporal rhythms that are part of the larger environmental system, it is nevertheless increasingly forced to operate under the standard discounting periods of corporate strategists and of economic cycles. Fifth, the privatized nature of crucial parts of the water cycle diminishes the transparency of decision-making procedures and limits access to data and information that could permit other social groups to acquire the relevant information on which to base views, decisions, and options. Finally, water production and distribution becomes incorporated into an increasingly global economy in which investment flows, financial capital markets, and

[3] However, we cannot dismiss the existence of powerful forces that oppose the privatizing agenda or the internal contradictions of the privatizing model, which has often ended in failure in many cases around the world (Savedoff and Spiller 1999; Hardoy and Schusterman 1999; Bond 1997).

investment decisions shape the contours within which the urban water economy operates. In sum, the shift from public good to private commodity alters the choreography of power through which the urban socio-hydrological cycle is organized (Kallis and Coccossis 2001).

2.3.2 The demand-supply-investment trialectic in a 'competitive' context

Within a context of commodification and demands for privatization, the traditional state-led way of managing the triad of demand-supply-investment decisions is fundamentally transformed. If the profit motive, either for public or private companies, becomes the yardstick against which performance is measured and the price signal a key instrument in regulating the demand/supply nexus, the contradictions between these moments in the economic process take a rather different turn. Investment is continually required to extend, replace, and update water supply networks. However, expanding demand in order to raise the necessary finance is seriously discouraged for environmental reasons, thus requiring more and more complicated equations on the balance sheets of water supply companies. With a given demand structure and the necessity of further investment, profitability (and hence the sustainability of market-led water companies) can only be maintained either via productivity increases (which are generally capital and technology intensive and almost invariably lead to a rising organic composition of capital) and/or price increases. When network expansion is required as in the case of most cities in the developing world, and usually in the poorer areas of the city, the substantial investment required hits the limited and often problematic cost recovery potential from investments in low-income areas. While price rises are possible (and likely), it remains politically sensitive and might lead to socially perverse effects. For example, immediately after privatization in the UK, the water price increased significantly. Many non-paying households were cut off (a practice later banned by the government), while companies and their shareholders gained considerable profits. In the second round of price-setting in 1999 (and after the government introduced a windfall tax on what were considered to be the excessive profits of the privatized utilities), price increases were more modest, immediately resulting in a major reduction of the labour force in the water industry and calls for a partial re-collectivization of the water infrastructure.

In a context of increasing demand and expansion of either total or per capita demand, the volume of profits can be maintained by means of an expansion of supply. In this context, it is interesting to note that the 'productivist' logic of water supply companies continues unabated. Furthermore, given the long-term and capital-intensive nature of investments in water infrastructure, there is a rather weak incentive to engage in major long-term and capital-intensive investment programmes. Put simply, there is a clear disincentive to invest in activities which are not directly profitable, such as leakage control and the expansion of the network, in contrast to productivity enhancing investments. Finally, in the context of geographically limited supply and demand in which most companies operate, while simultaneously being exposed to a rapidly internationalizing competitive environment, there is a tendency for privatized water companies to internationalize activities, either by taking over privatized water businesses elsewhere or by means of mergers, acquisitions, and/or diversification into other sectors (see Hall 1999).

In addition, the traditional state-led way of managing the triad of demand-supply-investment decisions becomes fundamentally transformed (see also below). If the profit motive, either for public or private companies, becomes the yardstick against which performance is measured and the price signal a key instrument in regulating the demand/supply nexus, the contradictions between these moments in the economic process take a rather different turn. In addition, new forms of governmental regulation parallel these shifts in the economic organization of water supply. This is the theme we shall turn to next.

2.3.3 A new regulatory order?

De-, re-, or non-regulation

The tendency towards commodification and privatization changes the regulatory context in important ways. While moves towards commodification and privatization are legitimated on the basis of considerations of increased competitiveness, higher productivity, lower prices, and drastic cutbacks of regulatory red tape, there has been a tendency to equate these shifts in the economic forms of organization with deregulation. However, evidence from the water sector suggests exactly the opposite. Particularly in the case of the UK, for example, the establishment of new semi-public institutions accompanied the privatization of the water utilities in 1989, most notably the economic regulatory body OFWAT (Office of the Water Regulator). Although the main function of OFWAT is the protection of the consumer by means of regulating price-setting and investment, this process proved to be full of tensions and conflict, not least as a result of a great and increasing diversity between water companies, uncertainties about available data, and the intricacies of the regulatory game. As Bakker (1999; 2001) has pointed out, the regulatory game that started with the privatization (and ostensibly deregulation) of the water industry unleashed a certain regulatory creep, which has developed into a top-heavy institutional regulatory body. Given the territorial monopoly character of the privatized water companies, all sorts of regulatory procedures, such as investment target-setting, pricing, environmental standards, abstraction and leakage standards, quality assurance, and the like, have been implemented. Rather than deregulating the water sector, privatization has resulted in a profound re-regulation of the water market and in a considerable quasi-governmental regulatory structure (Castro, Swyngedouw, and Kaïka 2003). In the process, the set of social actors involved in the institutional and regulatory framework of the water sector has been significantly altered, with a new geometry of social power evolving as a consequence. This new choreography of institutional and regulatory organization is what we shall turn to next.

The re-scaling of the governance of water: from water government to water governance

The host of new institutional or regulatory bodies that have been set up have considerable decision-making powers, but operate in a shady political arena with little accountability and only limited forms of democratic control (Guy, Graham, and Marvin 1996). These institutional changes have been invariably described as part of wider shift from government to governance (Swyngedouw, Page, and Kaïka 2002). Whereas in the past, water management and water policy were directly or indirectly under the control of a particular governmental scale, i.e. either at the national state and/or the local (municipal) level, in recent years there has been a massive proliferation of new water-related institutions, bodies, and actors that are involved in policy-making and strategic planning at a variety of geographical scales. The successive generations of water-related directives and regulations at the EU level, for example, or the tortuous process of implementing an integrated EU policy—in the form of the European Water Framework Directive—have resulted in growing powers of the Commission over water-related issues (Kaïka 2003*b*). In addition—as the UK case shows—privatization required setting up a series of new regulatory bodies (OFWAT in particular) and a redefinition of the powers and prerogatives of existing regulatory organizations such as the National Rivers Authority, which became integrated into the newly created Environment Agency. For cities in the developing world, international scrutiny and conditions, often imposed by global institutions like the World Bank or IMF, have significantly altered the choreography of power between national and international scales of governance. In addition, the negotiation, implementation, and follow-up of such arrangements are paralleled by a growing number of commissions, committees, and institutions, often of a public–private character, that supplant the traditional public authorities (Swyngedouw, Page, and Kaïka 2002; Castro, Swyngedouw and Kaïka 2003).

Finally, privatization itself of course results in much greater power and autonomy for the companies themselves, particularly in terms of strategic and other decision-making. Privatization *de facto* means taking away some control from the public sector and transferring it to the private sector. This not only changes decision-making procedures and strategic developments, but also affects less tangible elements such as access to information and data.

The combined outcome of the above has been a more or less significant reconfiguration of the scales of water governance. As Bob Jessop (1994) has pointed out for other domains of public life, the national scale has been redefined (and partially hollowed out) in terms of its political power, while supranational and sub-national institutions and forms of governance have become more important. Privatization, in turn, has led to the externalization of a series of command and control functions. The result is a new 'gestalt' of governance, characterized by a multi-scaled articulation of institutions and actors with varying degrees of power and authority, and in which traditional channels of democratic accountability are cut, curtailed, or redefined. This proliferation of 'governing bodies' at a variety of geographical scales has diminished the transparency of the decision-making process and renders it more difficult to disentangle and articulate the power geometries that shape decision-making outcomes. In practice, it can be argued that the transition from government to governance has implied—despite the multiplication of actors and institutions involved in water management—the transfer of key economic and political powers to the private component of the governance complex. This has not, however, happened in a social vacuum, but rather has fuelled a constellation of social and political conflicts, not least because of the consequences of an increasingly private-oriented governance model for the sustainability of socio-environmental systems.

The proliferation of regulatory bodies and systems of governance associated with the hydrosocial cycle at local, national, or international scales, has contributed to the emergence of a 'thick' regulatory structure, with ambiguously defined responsibilities and an imprecisely defined accountability. Different sets of actors are involved in the decision-making procedures depending on the geographical scale of organization or on the particular institutional embedding of the water companies. The choreography of such 'stakeholder' participation is uneven and unequal and, in many instances, operates outside traditional political democratic channels. While some actors are well represented in some settings, they are excluded from others; still others remain totally absent from the arenas of power where fundamental decisions are made.

2.3.4 Proliferating socio-spatial and socio-environmental water conflicts

The expanding scale of urban water operations as a result of either increasing per capita demand and/or a still growing urban population results in the need to continually expand the urban water footprint. In spite of attempts to manage demand, total production capacity continues to increase, resulting in either an effective growth of water extraction and/or growing pressures to expand water production capabilities. At the same time, alternative uses of the available water (ecological, recreational, industrial, or other) are encouraged, often in a context of extremely limited quantity or unreliable quality of available resources. Although pressures differ from country to country and from city to city, they are real and have led to more or less serious conflicts or threaten to do so in the near future. Perhaps the most notorious example of this is Tel Aviv. The existing national supply system from which the city draws its water is stretched to capacity in years of limited rainfall, and saline intrusion means that some aquifer sources can no longer be used. The now defunct peace process with the Palestinians has resulted in a promise to divert more water to Gaza where more than a million people live on currently a very limited supply of circa 25 litres per person per day. The negotiations with Syria on the future of the Golan Heights (parts of which are in the drainage basin of Lake Galilee, the country's most important water source), may also affect the total water balance of Israel. Negotiations are currently under way to buy and import water from Turkey, which has surplus water, partly as a result of the construction of the Anatolia water project, which captures headwaters from the rivers watering the Kurdish region and other Middle Eastern countries. If this goes ahead, a regional socio-spatial condition, which is already precariously balanced, will extend its geopolitical reach and intensify a complex and conflict-ridden situation.

In addition to these socio-environmental and spatial conflicts, the drive towards privatization has reopened the debate over the status of water. While general access to water at a very low or moderate price for the whole population was the received wisdom during the 'Statist' period, current practices aimed at running water services according to the market logic have reignited discussion about water accessibility. This is particularly acute in the developing world where growing numbers of urban dwellers are dependent on highly unreliable, unsafe, and costly systems of water haulage or delivery. The privatization of water businesses renders the prospect of expanding water services in large cities dependent on conditions of profitability. Needless to say, strategies of cherry-picking have been identified as private businesses seek out the most lucrative opportunities (Guy, Graham, and Marvin 1996; 1997).

The twin tensions of increasing the demand for urban water and the mounting pressure to allocate water to other functions have proliferated socio-spatial tensions and conflicts over water abstraction, water allocation, and water use (Crespo 2002*b*). These conflicts can take a variety of forms, including growing social differentiation within the city in terms of water consumption (see Chapter 3), conflicts over urban versus agricultural, industrial, or ecological use, and conflicts between resource extraction areas and urban consumption areas (reflected in conflicts over new reservoirs or dam constructions). In addition, the globalization of water companies signals a strategy in which local waters, turned into capital, are geographically reallocated to other places and cities. For example, London's water company, itself part of a global German conglomerate (RWE), has taken over part of Jakarta's water supply system. Invariably, the outcome of these struggles and conflicts is expressive of the uneven power relations that infuse the organization of the hydro-cycle.

2.3.5 The discourse of crisis: the contested politics of demand management

The discursive production of 'scarcity'

The risk of dwindling water resources, coupled with rising or unfulfilled demand, has intensified the political and social debate about the 'scarcity' of water (Nevarez 1996). As Kaïka's work has pointed out, this discursive build-up of a particular water narrative and ideology, which is particularly noticeable during conditions such as drought-related urban water crisis, serves specific political and economic objectives and policies (Kaïka 1999; 2003*a*). A climate of actual, pending, or imagined water crisis serves not only to instigate further investment in the expansion of the water-supply side (as in the case of Athens), but also fuels and underpins drives towards commodification. As the price signal is hailed as a prime mechanism to manage 'scarcity', the discursive construction of water as a 'scarce' good becomes an important part of a strategy of commodification, if not privatization (see Chapter 6). In this context, strange and often unholy political alliances are forged between free marketeers and parts of the environmental movement. The growing effectiveness of this movement in spreading the message of increasing (although socially constructed) water scarcity to the wider public can lead to greater willingness-to-pay, with a consequent acceptance of the market as the preferred (or indeed only) mechanism to allocate water resources. In addition, the discursive representation of water as being an integral part of nature permits casting 'nature' into pole position to explain scarcity. In other words, nature is the principal 'cause' of water scarcity rather than the particular political economic configurations through which water becomes urbanized in highly selective and socially uneven ways, resulting in a serious 'scarcity' for the poor and powerless and abundant waters for the socio-economic and political elites.

The politics of the technological fix

The management of the urban hydrosocial cycle, particularly the management of demand, operates largely via a combination of campaigns aimed at raising public awareness about water savings on the one hand, and attempts at reducing water consumption by means of a variety of technological fixes on the other. Generally the cost effectiveness of water-saving devices depends both on the price of the technology and the price of water. In a context of low water prices, water-saving devices are often not cost-effective. Although the aggregate effect on water savings is still disputed (most studies indicate a slowdown in the growth of water demand, but not a reversal of upward trends), the technological fix for water-related problems requires significant investment. Privatized water companies remain reluctant to invest in such technologies (given the cost implication), while public subsidization might be seen as a subvention to the private sector (in the case of a privatized water sector) or run against the dominant ideology of full cost recovery (in the case of public companies). Despite the availability of a wide range of water-saving devices and technologies, uptake remains limited and is not likely to have a major impact in the near future. More importantly, the displacement effects are almost invariably completely ignored and not part of the environmental audit, yet it is abundantly clear that environment-friendly technologies applied in one sector might have adverse effects in terms of the environmental effects of their own production process. A total environmental audit would be required in order to assess the net environmental benefit derived from a technological fix.

In addition, the master engineering narratives prioritize large-scale and centrally organized water supply systems at the expense of more decentralized and localized production and delivery systems, although the latter are generally the ones that serve the poorer residents of Third World cities. Given the marginal official interest in optimizing such systems, they remain either poorly organized and/or controlled by informal actors operating in a grey zone and providing services at highly inflated prices (see Chapter 7).

2.3.6 Globalizing H_2O and uneven development

The commodification and privatization of H_2O is increasingly embedded in processes of economic globalization. Whether publicly or privately owned, water businesses are expanding their operations geographically and becoming increasingly embedded in an international competitive process (Kazemir, Leinin, and Schaub 2002). In the case of privatized companies, furthermore, their capital structure is also becoming increasingly internationalized. For example, after the UK government opened the water sector to market competition in December 1994, a frenzy of merger and takeover activity started to take place. Many UK water companies are actively acquiring water operations elsewhere in the world, while British companies have been subject to takeovers from foreign competitors. For instance, Thames Water (London's water supply company) was acquired in September 2000 by the German multi-utility RWE. At a global scale, an accelerated process of concentration and consolidation is taking place that is rapidly leading to a fairly oligopolistic economic structure of water utility companies on a world scale. Regardless of the difficulties of regulating global companies (particularly with respect to environmental and social standards, investments, maintenance, and infrastructure upkeep), it raises the spectre of increasing geographical strategies around investments and about the spread of activities, the flow of water capital, and the portfolio of holdings. In addition, it opens up the possibility of strategic withdrawal of water companies from particular places and sites, permits strategic cherry-picking, and even the potential for bankruptcy or liquidation of activities. For a sensitive and vital sector like urban water supply, each of the above might potentially threaten urban sustainability conditions. In addition, it might lead to a situation in which the necessary provision of water for more problematic (i.e. costly) areas of the city has to be undertaken by the public sector, while the private sector picks places that optimize corporate profitability.

To the extent that water companies operate increasingly as private economic actors, they are also increasingly subject to standard market risks. While providing a fundamental and essential service, the economic survival of water operations is not necessarily guaranteed. Takeovers, disinvestments, geographical reallocation, bankruptcies, inefficient operations, and the like are of course endemic to a private market economy. In fact, this is exactly what market dynamics are supposed to do: weed out underperforming companies and reallocate economic resources from less to more profitable activities. This raises particular questions with respect to the long-term sustainability of market-based urban water supply systems. In the absence of strong incentives to enhance productivity or efficiency, and given the high cost and long time horizon of fixed capital investments in water infrastructure, private companies may fail to keep water systems running efficiently. This would, in the medium term, lead to a situation in which the State (at whatever level) has to once again become involved in the water sector in more direct ways. There is a tendency to leave the network/infrastructure part of urban water networks to the public sector, while private companies secure profitable management and operational activities. This entails an indirect subsidy of the private sector by the state and, in market terms, distorts the operation of the market. In a context in which the risk of water supply failure is too dramatic to contemplate, the state will have to remain (or become again) a key player in organizing water supply systems. This will become even more pronounced as environmental and sanitary standards in urban areas continue to decline.

2.4 Conclusion: the urban water conundrum

One of the most striking features of urban life is the ubiquitous necessity for (metabolized) water of a certain quality and quantity to sustain urban life and its fabric. Moreover, this water circulates through an intricate centrally controlled system to every single location in the city. At the same time, cities are confronted with huge volumes of sewage water, which is not only a problem in terms of its physical characteristics but also presents a serious health threat. This waste has to be removed again from every urban location by means of a similarly centralized and extensive sewage network. Every form of urban life depends on water but is simultaneously threatened by it. Therefore, it should not be a surprise that in the practices of everyday city life, water is a crucial material and symbolic good, which is embedded in and engenders urban social conflicts and struggles over its use and control. The realm of urban water, particularly under conditions of exclusion and problematic access, is indeed highly contested terrain.

In short, the urbanization of water and the social, economic, and cultural processes associated with its domestication have brought access to and control of nature's water squarely into the realm of class, gender, and cultural differentiation and struggle. The commodification of water, in turn, has incorporated the circulation of water directly into the sphere of money circulation. This makes access to water dependent on positions of social power, both economically as well as in terms of gender and culture. Intricate power choreographies have characterized the organization and management of urban water systems over the twentieth century. At the beginning of the new century, we are again embarking on a major transformation of the political and economic landscape of water production and delivery, one that is replete with all manner of tensions and conflicts. Before we embark on excavating Guayaquil's urbanization process through the lens of the political ecology of its own hydrosocial circulation process, we turn first to considering the water condition in the Latin American city in general, and in the Andean region in particular.

3

Water, Power, and the Andean City: Situating Guayaquil

> . . . life for the poor in many Latin American cities [is] a risky proposition. They breathe polluted air, drink contaminated water, eat unsafe food, and live among the garbage. They are subjected to earthquakes, mudslides and floods from an early age and have limited access to health and education, no money and no work.
>
> Anton 1993: 140–1

3.1 Water exclusion and the Latin American city

There is no aggregate shortage of water in Latin America. The Amazon's output into the Atlantic Ocean is about 150,000 cubic metres per second and a whole host of smaller rivers—the Magdalena, Orinoco, San Francisco, Uruguay, and Usumacinta rivers, to name but a few—all carry more than 1,000 m^3/sec of water into the ocean at their outlets. In contrast, Buenos Aires, Mexico City, and São Paulo, the three largest cities in Latin America, consume around 50 to 80 m^3/second, clearly a very small amount when compared to total available regional water resources (Anton 1993: 163). However, Mexico City is situated in an extremely water-scarce area, and other cities such as São Paulo, Brasilia, Guatemala City, Quito, and Bogota are located far from plentiful sources of water. Elsewhere, though, large cities and abundant water sources are in close proximity, yet large parts of their population still suffer from a lack of clean, cheap, and convenient water, a situation of scarcity in the midst of abundance. This chapter will examine the problems faced by the urban poor in Latin America in accessing potable water, and will examine the problems associated with its delivery.

3.1.1 The size and nature of the problem

Although it contains some very arid areas such as the Atacama Desert, Latin America is a humid region. Until recently, water was regarded as an abundant resource, and justifiably so: Latin America's annual precipitation is 60% above

Table 3.1. Average municipal water consumption in Latin American cities

City	Water consumption (litres per capita per day)	
	Source: Anton	Source: World Bank
Buenos Aires	630	
Havana	500	100
Maracaibo	475	
Córdoba	435	
Guayaquil	429	261
San José	423	
Monterrey	404	
Mexico City	360–527	
Lima-Callao	359	211
Curitiba	345	
Medellín	340	
Guadalajara	314	
Bogotá	304	
Santiago	300–555	286
Caracas	300–388	
Montevideo	289	
Quito	286–301	
São Paulo	270–293	
Salvador	266	186
Belo Horizonte	261	
Cali	237	
La Paz	177	73
Rio de Janeiro	188–684	299
Asunción	160–350	236
Barranquilla	148	
Cochabamba	130[a]	43

Sources: Anton (1993: 156); World Bank (1998: 278–9); [a] Crespo (2002*a*: 122).

the world average and the average annual run-off of 370,000 m^3 is 30% of the world total (Biswas 1979: 16). A glance at water consumption levels in Latin American cities indicates no aggregate shortage of water. Table 3.1 suggests that average daily water consumption in Latin America's big cities is comparable with that of cities in the developed world, and significantly higher than is the case in African and some Asian cities.

Given that the very minimum amount of water deemed necessary to sustain life has been estimated at 5 litres per capita per day (LCD) (World Bank 1976), and that under most circumstances 30/40 LCD is deemed sufficient for a reasonable level of personal and community health (Kirke and Arthur 1987: 125), even the city with the lowest consumption level would appear to have a plentiful supply of water. This impression is further reinforced when one considers that it has been calculated (Kalbermatten 1980) that neither personal hygiene nor public health requires water for domestic consumption to exceed 100 LCD. With this quantity of water, the full health benefits of a reliable water supply can be enjoyed and a water-borne sewerage system can be operated. Furthermore, domestic consumption beyond 100 LCD is thought to have little additional benefit to human health or environmental well-being. The inevitable conclusion is that, in all the above-mentioned Latin American cities, there is more than enough water to provide every member of the population with satisfactory water and sewerage systems.

Whereas in 1980 just over 200 million urban dwellers worldwide were deprived of water supply services, by the year 2000 this number had more than doubled to 450 million. For urban sanitation services the figures are respectively 295 million and 720 million (United Nations Centre for Human Settlements, various years). About a billion people suffer from chronic water shortages. The consequences of deficiencies in safe water supply for health and the environment are far more critical in densely populated urban areas than in rural conditions where there is often some reliable source of water available and waste is diluted more easily.

The physical expansion of Latin American cities during the past few decades required a parallel expansion of urban services. However, the sharpening of social and economic inequalities, combined with the institutional contradictions of water utilities (see Chapter 6), resulted in a spatial segregation process related to the resurgence of slums and marginal settlements in suburban areas of the city. Poor residents became systematically excluded from many of the basic services, including access to piped potable water. In the Metropolitan Area of Buenos Aires (El Gran Buenos Aires), for example, the proportion of people without water services rose from 6% in 1947 (out of 4.7 million inhabitants) to 24% in 1960 (6.7 million) and to 36% in 1987 (*c.*10 million) (Brustein 1991: 96; 1990: 190). As Table 3.2 suggests, up to 70% of the urban population in Latin America does not have proper sewage systems available, while up to a half (and sometimes more) lack relatively easy access to potable water (in the house, yard, or neighbourhood). Between 1980 and 1986, the urban population in Latin America grew from 224.1 to 275.1 million (from 65% to 69% of the total population), while the number of city dwellers deprived of easy access to potable water rose from 37.8 to 45.6 million (Saenz 1988: 2–3). Table 3.3 summarizes urban water access for selected cities in Latin America.

The exclusion of large segments of the urban poor from direct access to water expresses and unleashes a social, political, and economic struggle within the urban arena for control over and access to water. In general, there is no alternative source of potable water, so this has to be brought to settlements by means other than pipes. In many cases, private water vendors, who hold a *de facto* monopolistic control over this precious commodity, truck water in. In other cases, standpipes, wells, and/or long haulage journeys (customarily by women and children) provide some sort of access. The exclusionary practices

Table 3.2. Urban populations with access to water supply and sewerage in Andean and selected Latin American countries (percentages)

Country	Sewerage connections			Potable water			
	1987[d]	1990[c]	2000[c]	1988[d]	1990[c]	1996[e]	2000[c]
Bolivia	23	73	86	42	91	na	93
Colombia	61	96	96	92	95	88	98
Ecuador	36	na	70	58	na	82	81
Peru	55	77	79	52	84	na	87
Venezuela	60	na	71	90	na	na	88
Mexico	64.5[a]	na	na	81.5[a]	92	91	94
Argentina	79[b]	na	85	70[b]	na	71	85
Paraguay	88[b]	96	94	82[b]	85	70	95

Sources: [a] Coplain (1988: 9); [b] World Bank (1987); [c] UNEP (2002) and UNCHS (2001); [d] Vásconez (1991: 51); [e] World Bank (2000).

Table 3.3. Percentage of houses with indoor piped water and sewerage connections, selected Latin American cities

City	Source	Year of data	Percentage water	Percentage sewerage
Cochabamba, Bolivia	a	1997	80.7	—
	b	1993	33	20
	d	2000	57	—
Barranquilia, Colombia	a	1993	93.4	—
Santo Domingo, Dominican Republic	a	1993	86.8	—
Managua, Nicaragua	a	1998	58.4	—
Panama, Panama	a	1990	81.7	—
Guatemala City, Guatemala	b	1993	52	—
Recife, Brazil	b	1993	79	38
San Salvador, El Salvador	b	1993	86	80
La Paz, Bolivia	b	1993	55	58
Lima, Peru	b	1993	70	69
Asuncion, Paraguay	b	1993	58	10
Guayaquil, Ecuador	c	1990	64.0	55.2
	b	1993	80	55

Sources: (a) UNCHS (2001: 323); (b) World Bank (1998: 278–9); (c) INEC, Census 1990; (d) Crespo (2002*a*).

around urban water can be illustrated by the role of urban water vendors and the economic power they command by virtue of their power over a vital commodity. Monopoly rents (see Harvey 1974) can be extracted and appropriated through the mobilization of a geographically 'located' resource, which needs to be spatially 'relocated'. Their command over the spatial circulation of water

Table 3.4. Relationships between proportion of water consumed and percentage of households, and total water production per capita in selected Latin American cities

City	Percentage of population	Percentage of water received	Water produced (litres per capita per day)
Mexico City[a]	3	60	386.2
	50	3	
Guayaquil[b]	40	3	220.0
Lima[c]	43	88	368
	32	10.1	
	25	1.9	
Barranquilla[d]	30	5	
Santiago[e]	19	38	226
	19.4	9.1	
Cochabamba[f]	27	50	

Sources: [a] Illich (1986: 1); [b] Field work; [c] Espinoza (1988: 4); [d] Bernal (1991: 153–4); [e] Calculated on the basis of Icaza and Rodriguez (1988: 62); [f] Crespo (2002*a*: 121).

permits not only rapid money accumulation, but also gives the 'water speculators' a powerful position in the political economy of the city as urban life would be seriously, if not terminally, disturbed if water distribution were to come to a halt (see Chapter 7).

The problem, therefore, is not one of absolute scarcity but one of unfair distribution. Indeed, the failings of the water supply system to bring water to urban residents and their consequent dependence on alternative supply systems raises the question of the socio-spatial distribution of urban water or, in other words, of who gets what share of the available water. Even for those who are connected to the urban network, there is significant difference in both the quantity and the quality of water that can be accessed. Water pressure in the network is spatially often very uneven and becomes lower with increased distance from the central reservoirs. This leads to a condition of very irregular water supply, usually limited to just a few hours daily. Moreover, communities with low supply are usually found in poorer suburban settlements (Brustein 1988*b*). As the pipes are empty and air-filled during long periods of time, the danger for both bacteriological and physical (for example, corrosion) contamination increases and, consequently, the water quality is significantly lower and often becomes unsafe to drink.

In addition, there is a clear socio-spatial segregation in terms of use of water. Rosenfeld (1991: 187) maintains that in the case of Santiago de Chile (which nevertheless has a respectable 97.5% coverage), low-income residents consume on average 100 litres of water per day per capita, while high-income neighbourhoods show a daily per capita consumption of up to 800 litres. Table 3.4 shows the inequality in water distribution for a number of cities for which those data are available, clearly indicating that a tiny minority of urban residents consume the bulk of available potable water, while many have to make do with just a fraction of this.

In Guayaquil, 60% of the population consumes 97% of the produced potable water, whereas the other 40% has to make do with 3% of the available supply (Swyngedouw 1994). For those dependent on 'tanqueros', average daily consumption can accurately be estimated at 20 LCD. In Mexico City, 3% of the population receives 60% of the water, whilst 50% receives just 3% (Illich 1986: 1); in Lima, 43% consume 88% of the water (Espinoza and Oliden 1988: 4); while in Barranquilla 30% have to survive on 5% of the water (Bernal 1991: 153–4). Even in Santiago, one of the few Latin American cities which services over 95% of the residents with piped potable water, there is still a significant disparity: the top 19% of the population receives 38% of the water (Icaza and Rodriguez 1988: 62).

On a world scale, it is estimated (Christmas and De Roy 1991) that 1–2 billion people do not have access to a safe and reliable water supply. In short, water provision has increased living standards for those who have benefited, but around 20% of the human population has suffered from the exclusionary implications of water supply management approaches and still has no satisfactory access.

3.1.2 The consequences of insufficient water supplies for Latin America's urban poor

A deficient water supply can affect human health in a number of ways, causing diseases that are water-borne, water-washed, water-based, or water-related. Water-borne diseases are infectious diseases spread through water supply systems, water-washed diseases are caused by a lack of personal hygiene, aquatic invertebrate animals transmit water-based diseases, and insects dependent on water spread water-related diseases. The 1990s have seen cholera epidemics—beginning in Peru, where 200,000 cases were reported (Anton 1993: 162)—and spreading to neighbouring Andean countries, including Ecuador. Cholera is now rife in Amazonia, and has spread north to Mexico and south to Argentina. Yet the few cities with adequate water supply and sewerage systems serving virtually 100% of their population, such as Montevideo, have been little affected. Clearly, a well-functioning and comprehensive water supply and sewerage system is essential if a city is to resist the onset of not only cholera but also amoebiasis, diarrhoea, typhus, and hepatitis.

Quite apart from the obvious threat to human health posed by the lack of piped, potable water in many poorer neighbourhoods in Latin American cities, the networks of water pipes are often themselves the source of contamination. When, due to overconsumption in central districts, repairs, breakdowns, or insufficient input, pressure drops in sections of the pipes, supply to consumers is interrupted. It is an indication of the way in which socio-political interest groups have forged the delivery of natural resources to the human population that when this occurs, it is almost always the more peripheral, poorer neighbourhoods that receive the limited supply. Networks are often of a herringbone structure radiating from the city centre or reservoirs, so that while city centre fountains present an image of plenty, even the few urban poor who are connected to the system receive water for only a few hours each day. While this compounds the perilous position of the urban poor in relation to health, the danger of exposure to disease is further enhanced when pressure in the pipes becomes very low or even negative. Air fills the pipes, and contaminated water from the soil surrounding the pipes can often be drawn in, introducing another source of contamination. In São Paulo, this is a particular problem, as water is provided on a by-turns system, where each part of the city receives water for a certain number of hours, and then has no water for another fixed length of time. With the pipelines routinely drying up several times a week, the scope for contamination is alarming (Jacobi 1995: 13).

However, protection from ill health is far from the only advantage afforded by access to potable water. Where the piped water network does not service neighbourhoods, the process of obtaining water can consume large amounts of time and energy. Heavy containers may have to be carried from trucks, wells, or streams; people may have to queue for water; and precious fuel may be used for boiling the water. In short, water can become one's master rather than one's slave. Children can be forced to miss school and women may be unable to enter employment because of the need to stay at home to await irregular deliveries. Thus:

> The provision of reliable, safe and convenient sources of potable water will not only reduce mortality and morbidity but will also release those now engaged in collecting water for more useful tasks. (Kirke and Arthur 1987: 123)

When considering the ways in which deficient water supplies impact upon the urban poor, it is important to note that the effects are far from gender-neutral. Instead the 'feminization of poverty' (Jordan and Wagner 1993) has become an acute reality. Women in Latin America are rarely consulted about their water supply needs by overwhelmingly male-dominated water supply institutions, despite the fact that projects, which have fully involved women, have been shown to bring significant health and lifestyle benefits and have been better maintained.

Women are disproportionately affected by water supply deficiencies as they are invariably the primary procurers and users of water and are also given the sole responsibility for waste management (hence putting them at a higher risk of exposure to disease). They are also increasingly accounting for a significant proportion of family wage income, so that time lost waiting or queuing for water inhibits their ability to provide for their families. Yet community decision-making and planning processes operate exclusionary practices towards women. If women were involved, Jordan and Wagner (1993) argue, then projects would be more successful as women have a vested interest in their success. Moreover participation might also heighten women's self-awareness, respect, and recognition as valued members of the community. Thus far, it appears that while poor provision of water has adversely affected women's lives, women's association with water procurement has adversely affected water provision. Male-dominated institutions have given priority to solving male-dominated problems, of which water supply is not seen to be one. Yet this is clearly a myopic approach: although women may be the procurers and disposers of water, water quantity and quality are fundamental to the physical and economic well-being of the entire community. Where proper water and sanitation are provided, economic activity can increase as a result of the extra time available to women, which can only benefit society at large. For example, when piped water was introduced to Panama City, the production of goods by women doubled almost overnight (UNCHS 1985).

3.1.3 Managing the supply of water in Latin American cities

Demands for water in Latin American cities have mushroomed since the 1950s (Postel 1992) as a result of the spatial expansion of such cities, rapidly rising populations, and the development of industry. Globally, industry is now responsible for around 25% of all water use, and although industries can dramatically reduce consumption by recycling and reusing water, such practices have yet to be adopted in Latin America (Shiklomanov 1990). Demand has also been further boosted by the abundance and low cost of supplies for the most central and wealthiest neighbourhoods of cities, where horticultural and ornamental water use is common. Very often, citywide levels of demand cannot be met by supply, not because of any aggregate water shortage, but because of the massive overconsumption by the commercial sector and by the wealthiest residential areas. In order to deliver better water to more people, a change of approach is needed. The capital-intensive gargantuan projects aimed at creating ever greater supply to match ever increasing demand should be de-prioritized, and instead of raping ever greater swathes of nature in this way, we should look within cities to find more water. Quite simply, we should learn to 'do more with less' (Postel 1992).

Postel calculates that Latin American cities could cut their water use by around 33% without sacrificing economic output or quality of life. Moreover, investments in water efficiency, recycling and reuse schemes have been found to yield more water per dollar than conventional projects. Despite this, institutions and policies in Latin American cities hinder such developments at present. Some officials and all water vendors have an interest in maintaining the status quo. Moreover, the need to find practical solutions to the water problem is often deemed less important than the need to be seen to be doing something, resulting in a bias towards grand projects. A mayor is more likely to be remembered and re-elected for building a new pumping station or aqueduct than for supplying low-volume lavatory tanks. In summary, the problems of supplying water to Latin American cities which have developed since the 1950s have had very little to do with absolute scarcity, and have instead been caused by a lack of properly trained professionals, by political influence on technical decisions, by excessive bureaucracy in management and supply institutions, and by corruption in administrative and political systems (Biswas and Kindler 1989).

Although the paucity of water provision in poorer neighbourhoods is inextricably linked to the socio-political manipulation of nature in the city, it should be conceded that many Latin American city slums are located in marginal areas, where water supply systems have to overcome severe engineering problems. Shanty settlements on steep slopes above cities are often above the level of storage reservoirs, thus requiring expensive pumping of water to provide piped water. Many other slums are located on flood plains where the installation of water mains and drainage is again difficult and costly. Yet, these invasion settlements often take place on marginal low-rent yielding lands and, in the case of Guayaquil, were actually organized by a clientelist patronage system. In turn, the marginal conditions of the land lead to excessive costs when public or collective services need to be constructed. However, it is also clear that when deciding which areas are to benefit from a supply of water, institutions tend to favour the most affluent or politically influential neighbourhoods. Such areas may not only have the political power to act against people or institutions who make decisions which are not in their interest, but will also be more vocal, better funded and better trained in the ways and means of lobbying and influencing decision-makers. Furthermore, institutions perceive that middle- and higher-income groups are likely to be much more reliable at paying their bills, despite the fact that poorer households without connections pay far higher bills for non-potable water than they would for the same volume of piped water, and that, with the availability of piped water, opportunities to increase income would also expand.

One factor hindering large-scale development of water supply systems in Latin America in the 1980s and 1990s was the shortage of capital. Before 1982, although there was never a plentiful supply of capital, credit could be obtained from funding agencies, Western banks, or other financial institutions. However, since the debt crisis became apparent in the 1980s, the supply of capital and consequently of potable water for up to 50 million inhabitants of Latin American cities has dried up. Across Latin America, massive foreign debts were accumulated during the 1970s by the mainly military governments, as a result of unaccountable and corrupt regimes being encouraged to take out large loans by foreign banks. Thus, Brazil's debt increased from US$10 billion to US$100 billion, and Mexico borrowed $100 million despite a ten-fold increase in its petroleum income. Virtually every other Latin American country similarly obtained massive levels of borrowed capital. The price of these debts is now being paid for by the urban poor in two ways: in the high tariffs they are forced to pay for tanker water as a result of the failure of previous regimes to invest borrowed capital productively in water supply schemes; and in the economic austerity forced upon them by World Bank and IMF structural adjustment programmes.

The structural adjustment programmes imposed upon Latin American governments are now compounding this situation. In order to reduce fiscal deficits, the IMF insists upon smaller, self-financing public services charging higher prices, abandoning subsidies, and implementing privatization schemes. Reduced expenditure results in even less maintenance being performed and even fewer new projects being considered, thus perpetuating the exclusion of the poor from potable water resources. Furthermore, even where poorer neighbourhoods do have access to water connections, price increases have often put the cost of water beyond the means of the urban poor, leaving them dependent upon the standpipes and water vendors once more.

It has already been noted that the water supply problem is not one of absolute scarcity but rather one of 'produced' scarcity, and it is therefore worthwhile considering who manages water and in what ways it is mismanaged. In Latin America, water provision has traditionally been an area of public intervention, based upon the legal classification of water as public or government property. Most frequently, though, the potential for a well-coordinated water provision strategy that such public ownership might seem to offer is not realized due to the fragmentation of responsibility for water between numerous different public institutions. In Argentina, Bolivia, Chile, Colombia, Guatemala, Paraguay, Nicaragua, Uruguay, and Venezuela a plethora of institutions were involved but none took overall responsibility (Biswas 1979: 30). In Brazil, Costa Rica, El Salvador, Panama, and Peru there are also numerous institutions but one institution has a coordinating role and takes ultimate responsibility. Only in four countries—Cuba, Ecuador, Honduras, and Mexico—is water management centralized in a single institution. In recent years, water privatization programmes, like that in Buenos Aires or the failed attempt in Cochabamba in Colombia, have gained currency and have been portrayed as the panacea to solve both the financial crisis of public supply systems and improving the service. While the former objective might be achieved, there is little evidence that service provision and coverage has improved significantly.

Although giving one institution control over and responsibility for water should encourage wiser investment and strategic planning, it by no means ensures it. Institutions in Latin America have frequently been characterized by inefficiency, political interference, and excessive bureaucracy, all of which have reduced their effectiveness. Water company funds have, for example, often been diverted into other sectors of government, so that revenues—where they are accrued—are rarely used to improve the network. Water companies have also often been used as sources of employment for supporters of the political elite. Most such appointments are administrative and command significant salaries, so that much needed funds are spent increasingly on salary payments.

Moreover, the ensuing bureaucracy can be self-perpetuating. New employees expand their positions by creating new rules and procedures in order to protect their own jobs, and in so doing gradually create further posts. The casualty, as ever, is the budget for investment in infrastructure.

Fundamentally, however, the institutions of water management do not only suffer from problems of inefficiency, corruption, bureaucracy, and divided responsibility: even when funds are available for investment they are used in highly inappropriate ways. Almost invariably, funds are still pumped into improving and expanding large-scale water distribution systems, despite the fact that they increasingly provide more water for those who already have it without significantly expanding overall coverage. Encouraging greater economy in water use and extending provision to unserviced areas are seen as secondary priorities, despite the fact that they are normally more cost-effective. Moreover, investment in additional capacity without extensions to the system invariably leads to increased consumption among middle- and upper-income groups. Given that water always becomes more expensive to provide as the amount provided increases, increasing consumption above existing levels for the middle and upper classes has negative financial consequences for the water companies, without benefiting those in need in any way. In fact, water companies can find themselves caught in a spiral of ever increasing investment when providing water to city centres and wealthier districts, which has to be maintained to placate the politically vocal. Once water supply has been increased, additional piped sewerage is required. As households become more accustomed to in-house water and install baths, showers, irrigation systems, and multiple taps, consumption increases still further. As a result of this, mains pressure drops and householders complain. Further investment is then demanded, and as a result of this extra water, the cycle begins again. Just as those who have power control the water, and those who have water manipulate those who have power, those who have no power have no water.

Crucial to any analysis of water supply and demand is an understanding that water demand depends very much upon availability. Where water is made widely available at low prices (as it is in most Latin American city centres and wealthier residential districts), consumption practices are highly wasteful. Awareness of the true value of water is minimal, water-saving technology is scarce, and pricing policies do not promote conservation. In Buenos Aires, for example, a combination of plentiful supply from the Rio de la Plata and a lack of metering have brought about extremely high consumption patterns. Water is often supplied at minimal cost—even below cost price when capital investments, maintenance, and other expenses are taken into account—in well-intentioned yet unsuccessful attempts to ensure that water prices do not put water beyond the means of the poorest. The actual results of this policy are that wealthier residents take access to large volumes of water at low cost for granted, and use so much artificially cheap water that the poorer, peripheral areas which do have connections to the piped network suffer minimal or no pressure and the districts without connections see funds which could be spent on supplying them wasted on subsidizing garden sprinklers and fountains for the wealthy. In Lima for example, 'leisure lakes' are fed by the municipal La Atarjea system, but more than 300,000 households have no service whatsoever (Anton 1993: 155).

Thus setting artificially low tariffs can be seen to be a pricing policy that has failed to benefit those in most need, and has in fact worked to their detriment. Clearly, a more efficient tariff structure is urgently needed if water agencies are to use their own resources properly and are to be able to extend their network. Such a structure should aim to provide low-cost water for essential domestic use but to charge high tariffs for any additional water use. For example, cheap water could be supplied up to a volume of 100 LCD, with additional water available at rapidly increasing tariffs. Such a system would require the existence and satisfactory operation of water meters, but would be far superior to the current situation in many Latin American cities where water use is charged at a flat rate or is even charged at decreasing unit costs with increasing consumption. There is no financial, engineering, or natural justification for such a tariff system: it is indeed a vivid illustration of the way in which the management of water supply is more a process of political management and manipulation than natural resource provision.

The above provides the background against which the political ecology of water urbanization in Guayaquil is framed. In the next section, we shall turn to presenting the water condition in the city. The subsequent chapters, then, will delve into the excavation of the political ecological processes through which the urbanization of water in Guayaquil unfolded.

3.2 Exclusionary water practices in Guayaquil

3.2.1 The geography of water exclusion

Guayaquil is the largest and economically most powerful city in Ecuador. Situated on the Pacific shore of the country's humid lowlands, it suffers from immense water problems. Billions of litres of water pass through the city centre every day as the Rivers Daule and Babahoyo join to form the Guayas stream, while almost half of its residents do not have access to reliable sources of potable water and the whole city suffers from chronic water shortages. While Quito is Ecuador's political centre and capital city, Guayaquil is the country's hustling and bustling port city, whose location is shown in Figure 3.1. Together with Duran, located on the other side of the Guayas River, Guayaquil's metropolitan area today includes approximately 2 million inhabitants. About 600,000 of these live in unregulated or poorly regulated settlements that grew out of invasions of landless rural workers, who started to migrate to the city from the 1950s onwards.

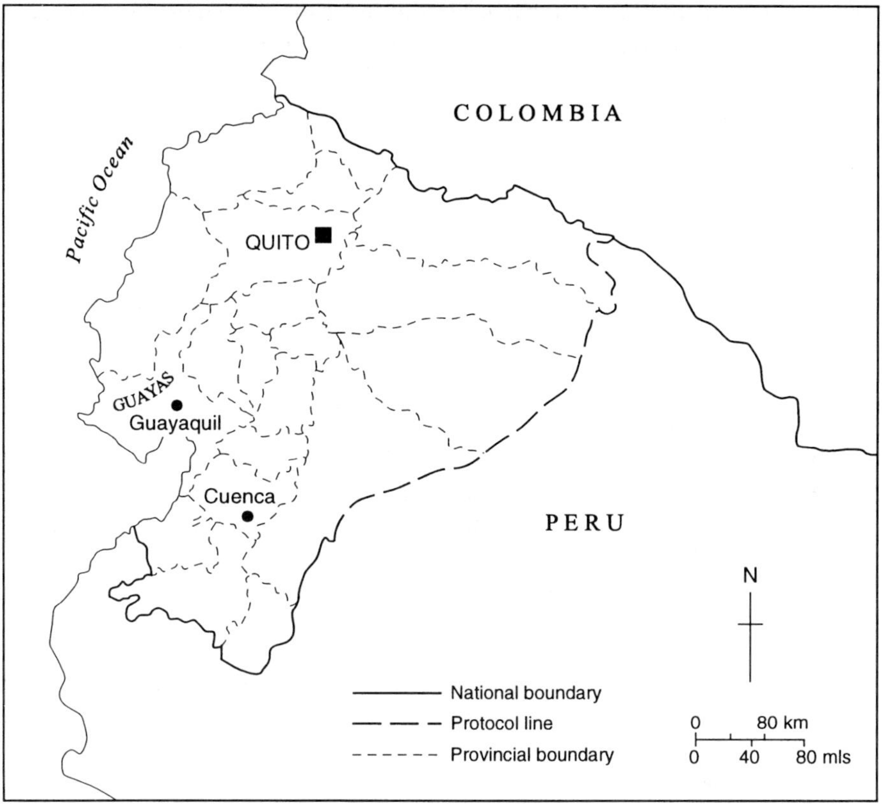

Fig. 3.1. The location of Guayaquil in Ecuador.

The growth of metropolitan Guayaquil ran increasingly ahead of the provision of water services. As the political-ecological transformations of the countryside disintegrated rural society and caused accelerating rural to urban migration, the state as the key locus for the provision of collective consumption equipment failed to appropriate the necessary rents from the ecological conquest of the urban hinterland to assure a parallel expansion of urban services (see Chapters 4 and 5). Table 3.5 summarizes the recent evolution of domesticated water in Ecuador, Quito, and Guayaquil. While the national average showed signs of improvement over the 1974–90 period, the situation in Guayaquil deteriorated significantly, both in relative terms and in absolute numbers. The rate of coverage fell by 9%, while the absolute and official number of city dwellers lacking access to piped water grew from 222,269 to 596,013, almost tripling in less than 20 years. Of the 169 cantonal capitals in the country, 144 enjoy a better service than Guayaquil and 114 do better than Quito.

Table 3.5. Potable water and sewerage services in Quito, Guayaquil, and Ecuador, 1974–1990 (percentage of dwellings connected)

	Quito	Guayaquil		Ecuador
		%	deficit	
Potable water				
1974	85	73	222,269	43.7
1982	85	65	419,770	51.8
1990	83.3	64	596,013	57.1
Sewerage				
1974	89	82		28.1
1982	—	52		33.6
1990	79.8	55.2		39.5

Sources: INEC, Census 1974; 1982; 1990.

Table 3.6. Water accessibility and water provision in the metropolitan area of Guayaquil (City of Guayaquil plus Duran), 1990

	Houses	%	Inhabitants	%
TOTAL	349,176	100	1,643,207	100
In-house	163,183	47	743,978	45
Outdoor	43,696	13	202,476	12
Neighbourhood	18,887	5	92,129	6
No-water	123,369	35	604,624	37
Public network	219,439	63	1,007,574	61
Private vendor	121,257	35	593,731	36
Well	4,315	1	21,315	1
River	1,410	0	7,031	0
Other	2,755	1	13,556	1
Sewerage	184,998	53	834,199	51
Collected waste	192,811	55	878,314	53

Source: INEC, Census 1990.

The 1990 census data for Guayaquil presented in Table 3.6 give further details of water accessibility and the means of water provision in the city. Only 45% of the urban residents enjoy the luxury of fully domesticated water flowing through indoor plumbing. A further 18% have some form of access to the official public water network, whereas the remainder are dependent on other means to acquire their necessary supply of water. The overwhelming majority of those who are excluded from the engineered water supply system rely on

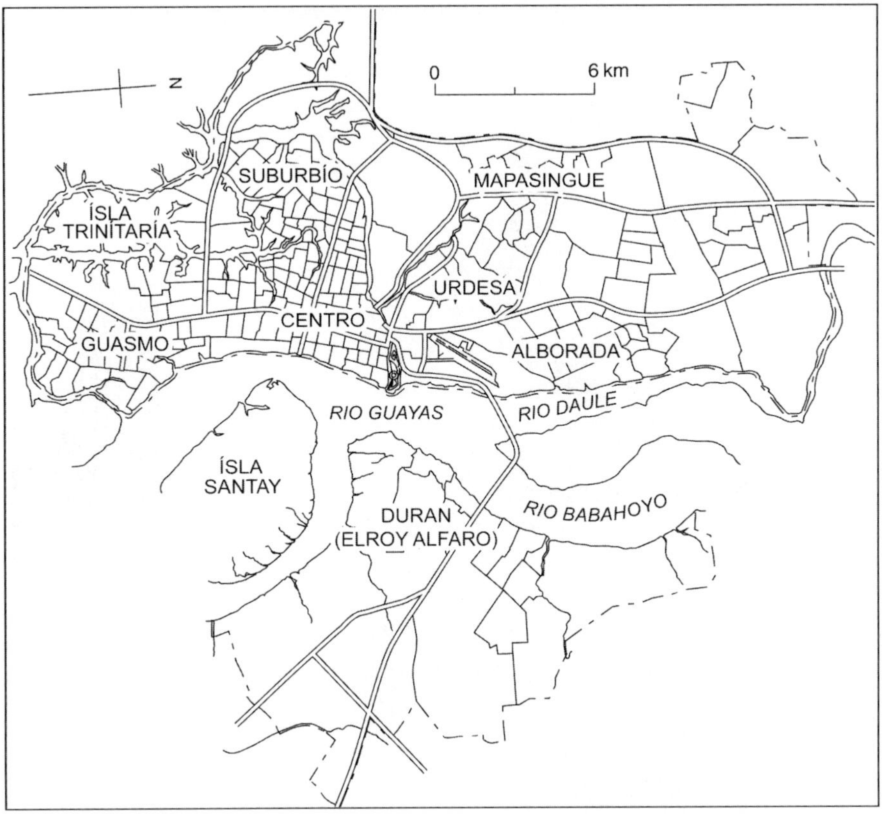

Fig. 3.2. The city of Guayaquil and its main urban divisions.

private water vendors. The number of those dependent on these monopolized water speculators has grown spectacularly over the past 20 years, from about 200,000 in 1974 to almost 600,000 in 1990. However, most experts agree that, on the whole, the census data underestimate the real figures, particularly in the marginal peripheral settlements. According to these observers (see Scheers 1991), the actual population in 1990 was closer to 1.8 million, which would make the number of people dependent on private water sellers closer to 800,000 than to the tabulated figure. Other sources, therefore, claim that the rate of water coverage is significantly poorer than assumed. Arellano (1992), for example, maintains that the actual rate of water coverage fell from 76% in 1975 to 54% in 1991 and to as low as 50.3% in 1992.

Figure 3.2 shows the location of the main urban areas of Guayaquil. Figure 3.3 details the geography of water exclusion. The settlements in the southern and northwestern peripheries are among the least serviced areas. In fact, there is a clear water gradient from the central parts of the city to the periphery. In

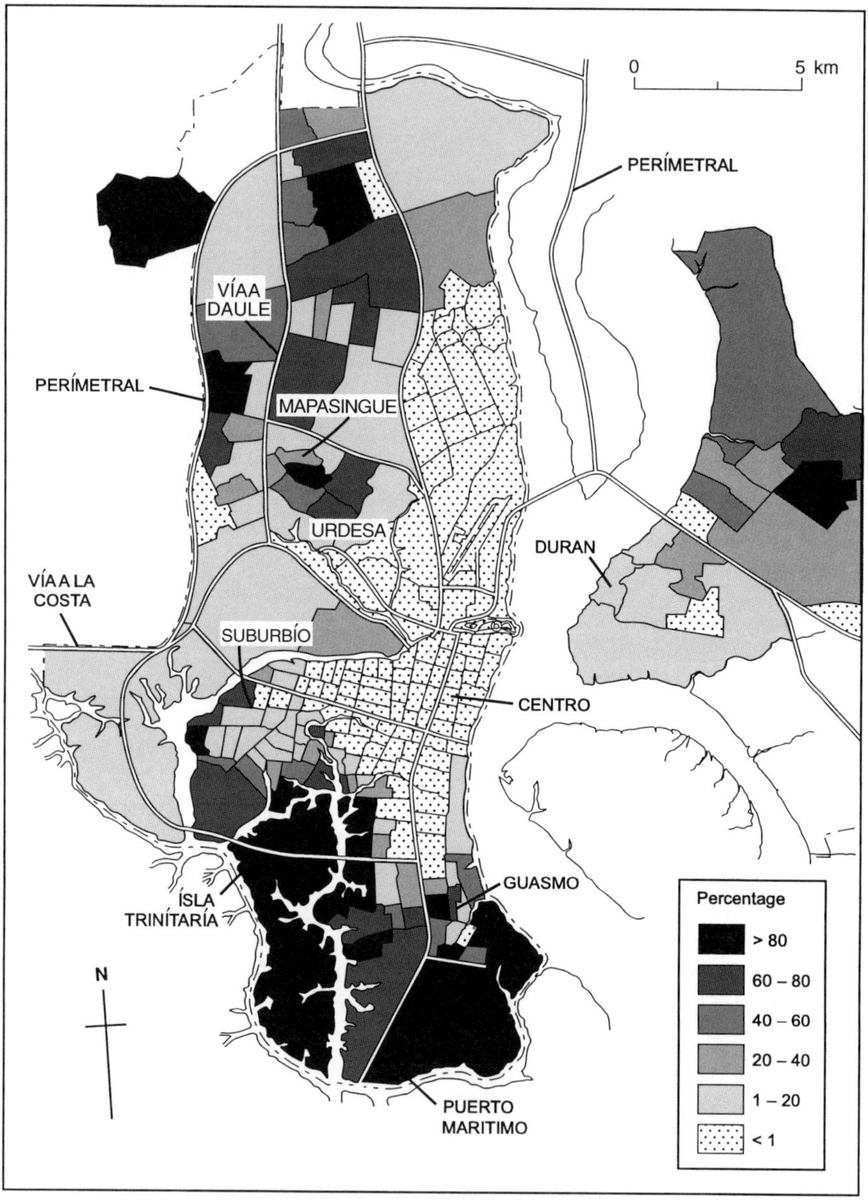

SOURCE: INEC, 1990.

Fig. 3.3. Percentage of dwellings served by water lorries in Guayaquil, 1990.

the most recent settlements (land invasions of the last 20 years), the whole population is dependent on private water sellers, while the middle- and upper-class residential areas to the north of the city centre are fully serviced through the public water authority network. These data do indicate the socio-spatial unevenness in terms of access to nature's water and suggest how the flow of water can indeed shed light on the mechanisms of socio-economic and political power that shape the urbanization process and give the city its highly contested, unequal, and oppressive characteristics.

Figure 3.4 shows the water supply system of Guyaquil. The pumping station at 'La Lolita', 95 km from the city, was the sole water supply system for the city until 1950, although the connection with Guayaquil was severed in 1975 and at present it only services Duran and the villages along the pipeline. In 1950, a pumping and treatment station was built 25 km upstream from Guayaquil on the Guayas River. This station now has a theoretical annual capacity of 1.5 million m^3 and mains pump the water to the city centre and to Salinas, a resort town 140 km from Guayaquil. Centrally located reservoirs (around which—not surprisingly—the middle- and upper-class residential areas are located) are then the nodal points from which secondary mains and pipes service the residential and central commercial areas along the model of a classical herringbone structure.

But network connections do not guarantee a steady supply of water. The technological structure of the system ensures high pressure and good quality close to the urban reservoirs, whilst further away pressure falls rapidly and supply is limited to a few hours a day. In fact, social inequality and water exclusion is cemented into the technological engineering system itself. The suburban areas, deprived of water connections or faced with chronic supply shortages, are dependent on private water sellers (tanqueros) for their essential supply of water. The latter procure their water from the public water company at three filling stations, located along the mains that bring the bulk water to the city (see Chapter 7). The northern sector of the city consumes on average 307 LCD, while the southern sector has to make do with only 43 litres (see Table 3.7). The northern part has an around-the-clock supply, while the southern sector receives water of dubious quality for only 4 hours a day. Moreover, the technical efficiency of the system is extremely low. In the areas with high pressure, more than 50% of the water is lost through leakages and as a result of insufficient accounting systems. Of course, the unreliable supply in the central and southern parts of the city also leads to a social stratification of water use. While average per capita consumption is well over 300 LCD, it is only 43 LCD in the peripheral areas of the water system. This is still double the volume consumed by those who are dependent on water vendors.

Nevertheless, the average production and supply capacity of the existing facilities allow for a daily per capita consumption of 220 litres. Compared with an international standard of 150 LCD, Guayaquil would be in a position to provide every citizen with a sufficient supply of water. The key issue, therefore,

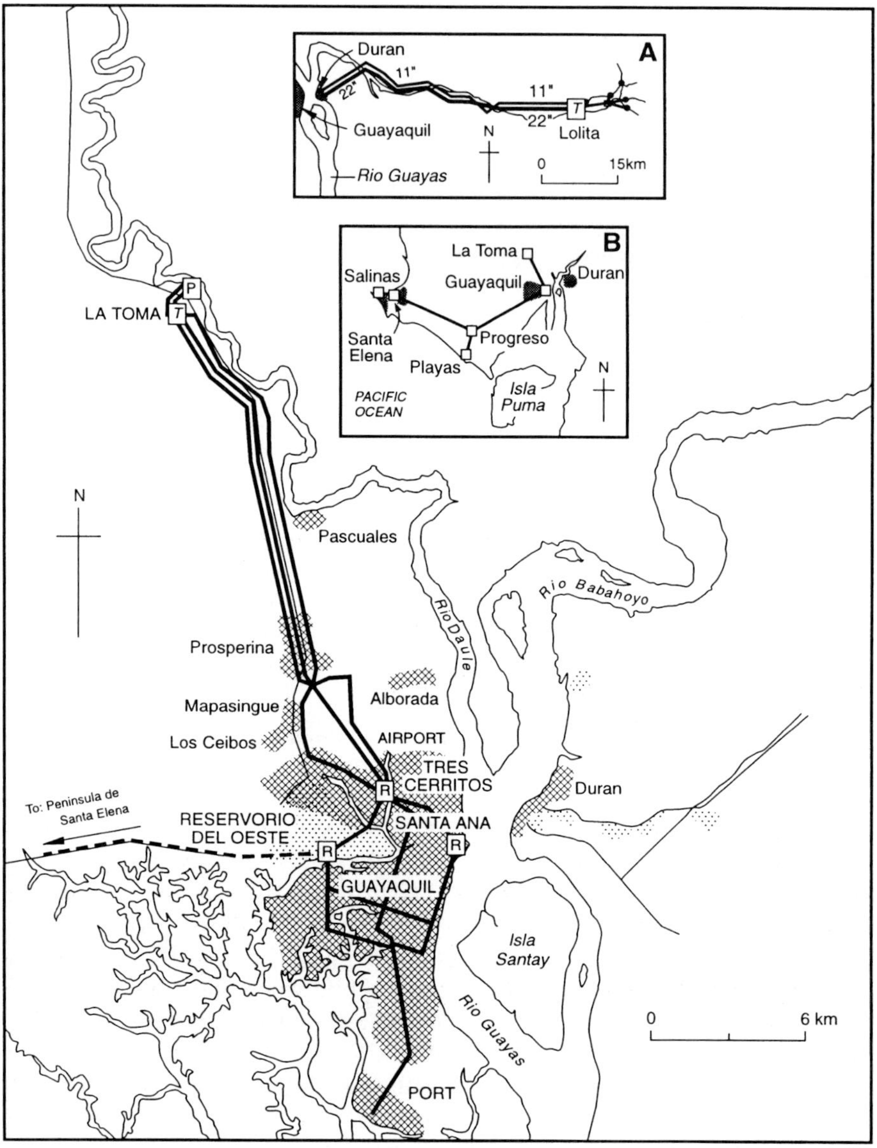

Fig. 3.4. The water supply system in Guayaquil.

is one of distribution of available capacity rather than an issue of absolute scarcity. The water scarcity experienced in some sectors of the city and among some of its residents is socio-politically constructed rather than produced by environmental or production constraints. In addition, supply problems within

Table 3.7. Geographical distribution of water supply and consumption through the official network, 1990

Sector	North	Centre	South
Number of inhabitants	421,214	422,985	272,393
Water supplied (m^3 litres/day)	272,471	99,500	16,000
Water/inhabitant/day (litres)	307	160	43
Average hours of service	24	10	4

Source: EPAP (1991*a*).

the piped network make the existence of relative scarcity more acceptable to large sections of the population—both poor and rich—and lower the expectations as to the quality and quantity of the services provided. As Vásconez (1988*b*: 14) puts it:

> In the consolidated [improved] invasion settlements, the problem of potable water is not entirely resolved, but at least the majority of the dwellings are connected to the supply network. Although the levels of supply are considered insufficient, the situation really becomes unsustainable only during periods of drought or rationing. Here, a phenomenon occurs that is the opposite of how abundance of supply makes shortages infamous and unjust. In fact, the perception of absolute shortages renders relative shortages acceptable to a certain extent. The existence of extreme poverty, therefore, leads to a lowering of expectations of the less poor. (my translation)[1]

In short, the ideology of production management, serving clear local interests and propagated by international financiers and official development agencies, produced and perpetuates the existing mechanism of uneven access and outright exclusion and obfuscates both the issue of a just and empowering distribution and the pressing problem of equitable water management. Ironically, this constructed scarcity lowers the expectations of many urban residents, and consequently helps to defuse the potential for social mobilization and grassroots rebellion. Nevertheless, the average daily production of water per capita does indeed suggest that sufficient (although certainly below Western consumption standards) amounts of water are available to allow for the whole of the urban population in these cities to have sufficient water for a healthy and acceptable standard of living.

[1] En los barrios populares consolidados el problema del agua potable no està enteramente resuelto, pero, por los menos, la mayoria de las viviendas està connectada a las redes y, si bien, la cantidad de usarios vuelven insuficientes las dotaciones actuales, las situaciones sòlo se vuelven insostenibles en épocas secas o de 'racionamiento'. Ocurre aqui un fenòmeno inverso al de las opulencias que habia notorias y injustas las carencias: al contrario, la percepciòn de las carencias absolutas, par contraste, vuelven en cierto modo y hasta cierto punto, soportables las situaciones de escasez relativa. La existencia de probrezas extremas puede, de hecho, empujar a la baja las aspiraciones de sectores menos pobres.

3.2.2 Urban segregation, land rent, and the control over water

The expansion of the city due to the exodus of rural peasants and rapid internal growth took place largely on marginal lands without infrastructure, with difficult topographies, and which yielded very little, if any, actual or prospective land rent (Rojas and Villavicencio 1988; Brustein 1988*c*). Vásconez (1988*b*: 7) argues that the spatial structuring of water provision, i.e. the geographical distribution of urban areas with different levels of access to water, resulted in a differential valorization of the price of land and definition of urban land use. This condition accentuated the housing problem of the poor as their economic position and the high price of (serviced) urban land forced them to locate in areas deprived of infrastructure and collective equipment (with the exception of the 'tugurized' central areas[2]) because the very absence of facilities kept land prices low (Brustein 1988*c*; Vásconez 1988*b*). Moreover, this process was often implicitly and explicitly encouraged by the local state as the occupation of marginal low-value land preserved and, in the end, increased the potential rent of developable urban land in prime suburban areas. As such, marginal suburbanization became an element in and an expression of wider urban land speculation on the one hand, while serving clientelist interests through first condoning and later actually organizing land invasions and the subsequent piecemeal provision of some basic services (roads, electricity, and in some cases water) on the other (see Rojas and Villavicencio (1988) for a discussion of Guayaquil). The invaded lands consequently became pivotal areas for cultivating a clientelist political system of patronage, both through the organized provision of land and the personalized delivery of services.

Ironically, the initial advantage of low urban rent values is quickly replaced by the extraction of extortionate geographic or location rent through the monopoly control of water vendors over the distribution of water. What was saved on land rent is spent many times over on the purchase of water (see Chapter 7). In this sense, the housing question is deeply related to the question of access to other key consumption commodities. While families could afford to move to under-serviced marginal lands exactly because the absence of services kept land prices down, other spatial rents resulting from inadequate or lacking public provision skyrocketed. The social production of nature and its recycling through monetary circulation expresses and creates political economic relations of power, domination, and exclusion. The urban expansion into the mangrove-covered areas south of city (which form Guasmo and Suburbio and now extend into Isla Trinitaria (see Plate 3.1)) and the land invasions, which form Mapasingue and Bastion Popular (see Plate 3.2), exemplify the above considerations for the case of Guayaquil (see Figure 3.2). Conversely, inhabitants of marginal urban settlements are often reluctant to

[2] 'Tugurizaciòn' refers to the process of central city densification through the internal division of houses into rented accommodation to house incoming rural families. Service provision is usually precarious and living conditions leave much to be desired (Cadme and Morocho 1980).

Plate 3.1. Drowning in water and starving from thirst: Isla Trinitaria, Guayaquil.

Plate 3.2. High and dry: Bastion Popular, Guayaquil.

struggle for domestic water supply connections not only because of their often precarious and uncertain legal and institutional positions, but also because service provisions are likely to cause an increase in land values. The latter might in the end even affect their positions within the area as higher-income groups, attracted by the improved servicing and increased value, might in fact force the original residents out. The poor are, consequently, caught between paying higher water rents or facing the consequences of increased land rents. A class gradient therefore emerges, in which consolidated and serviced formal invasion settlements are increasingly occupied by the middle classes, pushing poorer residents and new immigrants into the more peripheral and more recent invasion settlements.

Moreover, the specific geographical characteristics of marginal settlements—poor location, difficult topography, and obsolete infrastructure (in the case of 'Tugurized' city centres)—facilitate the continuing exclusion of the urban poor by reinforcing technical arguments and blaming the lack of investment funds as the main reasons for continuing water deprivation. The apparent technological managerial nature of the problem of urban service provision further feeds the productivist logic, which both friends and foes regard as the key issue. The operation of the political and socio-economic dynamics shaping peripheral urbanization and water distribution can be discarded or disguised in the 'ideology of underdevelopment', by blaming a lack of financing, expertise, and technology for the problems of difficult access and lack of supply. In addition, the 'bunching' of services in the central city areas which more or less coincides with the spatial extent of the water network produces a bundle of mutually reinforcing positive externality effects for the industrial, commercial, and financial activities which concentrate there, and further accentuates segregation and exclusionary urban spatial and residential organization.

Finally, in contrast with agricultural consumption, urban water use is characterized by the social and physical metabolism of water rather than its final consumptive use or its integration into new and transformed commodities. This means that used water remains in place as residential, commercial, or industrial waste water and poses a serious health threat if not efficiently removed. As the data in Tables 3.5 and 3.6 show, infrastructure for the second half of the circulation of water, the physical evacuation of sewage from the city, trails even further behind potable water supply systems in terms of area covered. Evidently, the absence of adequate sewerage systems to deal with the millions of cubic litres of more or less contaminated urban water is most acutely felt in the marginal suburban settlements and coincides with the absence of potable water supply networks. If we add inadequate garbage collection to that list (see, for example, Olaya 1991), the marginal settlement dwellers are literally trapped and drowning in their own excrements without access to water to clean it up or circulating conduits to flush the houses and streets. At the same time however, the sharpening of these segregation process and ensuing social conflicts intensifies the urban crisis and contributes to the further deterioration of everyday urban life. Even the rich can only escape from this urban degeneration by exercising their superior command over space, and catching the next plane to Miami.

3.2.3 *A matter of life and death: the geography of the hydraulic health problem*

The social struggle for control over and use of available water is clearly not only unfolding within the urban arena with its characteristic mechanisms of access to and exclusion from water, but also takes a much wider form in the socio-geographical rivalry between alternative uses, i.e. agricultural, industrial/commercial, and residential. Not only has groundwater and fluvial pollution increased over recent years as a result of seepage and disposal of heavy metals, synthetic and agricultural chemicals, and other hazardous wastes, but over-pumping has also caused salt water to infiltrate freshwater aquifers or to move further up river mouths. This, combined with malfunctioning or absent sewerage systems, negatively affects water quality, poses serious health threats, and pushes water treatment costs spiralling upward (World Bank 1992: 47). For example, in Lima upstream pollution has increased treatment costs by about 30% (World Bank 1992: 101). In Guayaquil, upstream irrigation projects not only divert water from the River Daule (an estimated future use of 100 litres/second), but are also associated with more intensive agriculture and its associated fertilizer-rich run-off, leading to potential eutrophication problems in the river. This has already led to regular spectacular blooms of water lilies, which, in turn, clog the filters and diminish the pumping capacity of the water treatment station. Periods of fast growth invariably lead to water shortages in the city. This process is also likely to cause higher water production costs in the future.

Within the urban sphere, the struggle to acquire water and remove sewage is quite literally a matter of life and death. There has been a resurgence of cholera epidemics in Latin America (particularly in the Andean countries of Peru, Bolivia, and Ecuador) since 1990, and the disease is now endemic in some regions. This is the most dramatic example of the deterioration of sanitary conditions in urban slums and is largely caused by ineffective sanitation, poor water supply, and insufficient or non-existent sewerage connections. For example, in the province of Guayas, there were 14,951 reported Cholera cases (5.1/1,000 inhabitants) in 1991 and a further 11,558 in the first ten months of 1992. In these two years, cholera was the third most frequently reported illness on a list of 40 diseases, after influenza and diarrhoea (itself related to water quality). Moreover, of the 50 main causes of mortality, gastrointestinal diseases (mostly related to unsatisfactory sanitary and hygienic conditions related to water quality and waste-water disposal) came eighth for the city of Guayaquil (14.3/100,000 inhabitants). In the marginal urban areas, drinking water was in 1991 the fourth most lethal activity (after heart diseases, pneumonia, and traffic accidents, but before homicides) (Dirección de Salud de Guyas 1992). Water-related diseases were directly responsible for 3.5% of all deaths

Table 3.8. Bacteriological analysis of drinking water in selected parts of the city

Location	Total count/ml.	Coliforms/ml.
Av. Carlos Julio Arosemana	5	0
Av. De las Americas	0	0
Las Peñas	3	0
Centro de Guayaquil	13	0
Av. Domingo Comín (sample 1)	Uncountable	10
Av. Domingo Comín (sample 2)	30	0
Guasmo	Uncountable	0

Note: The bottom three are samples from water sold by 'tanqueros'.
Source: Fundación Natura (*El Universo*, 31 July 1992).

for the city as a whole and for 4.9% of deaths in the peripheral settlements (Suburbio, Guasmo, Isla Trinitaria, Mapasingue). Both the illegally tapped water as well as the water distributed by water vendors is often highly bacteriologically contaminated, with serious negative health consequences. Table 3.8 shows the level of bacteriological contamination of drinking water in a number of areas in Guayaquil.

For children, the water-related health problems take dramatic proportions. Intestinal infections are the fourth most important cause of infant mortality (3.3 per 1,000 live births in 1992) and are by far the most important infant illness. In an average week in Guayaquil (data for 24–31 October 1992), diarrhoea was by far the most reported disease among children under 5 years old on a list of 22 controlled epidemiological illnesses. Of a total of 1,850 cases, 1,018 were diarrhoea-type diseases (55%), followed by 613 cases (33.1%) suffering from sub-optimal weight and malnutrition (Dirección de Salud de Guayas 1992).

In addition to the evident health problems and the uneven geography of water-related deaths, the struggle over water is also related to forms of social domination and exclusion and to psychological stress. The uncertainty about daily water supply, the time invested (usually women's time), and the energy spent on waiting, hauling, carrying, or working for water, waste productive time and energy. Above all, however, it restricts the time available for creative and emancipatory living as the exclusionary practices of the political economy of water control submerge other activities under the daily quest for urban survival.

Moreover, while deeply implicated in the reproduction of labour and family relations, the struggle to get water simultaneously threatens this very process. In addition, the search for water is by no means gender-blind. On the contrary, the struggle for water reinforces gender divisions and gender domination. For example, private water vendors are, without exception, male, while the overwhelming majority of buyers and main users are female. The uncertainty and irregularity of water supply and its often time-consuming character forces one family member, again usually the woman, to stay close to the house to attend to the daily routine of water buying and/or haulage. The high cost of water is compounded with the loss of time that could otherwise be spent more productively. In addition, the water is often of poor quality and unsuitable for drinking, thereby increasing expenditure on alternative, but equally commodified, substitute liquids such as sodas (Coca-Cola is ubiquitous, but its price is equal to about two hours of labour for a minimum wage earner), mineral, or purified water. The latter is usually the cheapest alternative at approximately US$0.5 for 4 litres.

All of this suggests how the circulation of water is caught into the contradictory development of the urbanization process and the political economy of power that shapes the socio-spatial structure of the city. Indeed, the serious shortcomings of water provision in Guayaquil bring water to centre stage in the relationships of political, economic, and cultural power through which the urbanization process takes place. The struggle for water and the contested nature of the uneven access to water turns the water issue into a highly contested terrain. Elements of class, gender, and ethnic relationships become embedded in the circulation of water. The water circulation process, therefore, can be used as a means to excavate the multiplicity of power relationships within the city.

However, before we turn to the political ecological processes through which access to and exclusion from water is organized, it is imperative to reconstruct the historical geographical processes through which the urbanization of water in Guayaquil was organized. In the following two chapters, the history of Guayaquil's urbanization process will be written from the perspective of the need to urbanize and domesticate nature's water and the parallel necessity to push the ecological frontier outward as the city expanded.

PART II

Social Power and the Urbanization of Water in Guayaquil, Ecuador

4

The Urban Conquest of Water in Guayaquil, 1880–1945: Cocoa and the Urban Water Dream

> About 40 miles from Duran we passed the water purifying plant of the Guayaquil water system. The source of this supply is in the high mountains in the backcountry. Before I left home an old sea captain warned me not to drink water in Guayaquil unless it had been boiled . . . I am sure now that the Captain sailed the seas before the Rockefeller Foundation made Guayaquil sanitary and in so doing wiped out yellow fever for which it was notorious, and before the city installed its present modern water system.
>
> Desmarest 1937: 54

> Tratemos de imaginar, por un momento, las soleadas calles del centro del puerto recorridas—de tanto en tanto—por un parsimonioso tranvía eléctrico o un lustroso Ford, o transitadas por unos cuantos peatones mostrando impecable levita y sombrero 'tostada', accompañado damas de botín, falda al pie y sombrero con encaje; al cine Edén iban por las tardes a espectas—antes que en Buenos Aires o Santiago—los films por rollos de los Barrymore, mientras el Teatro Olmedo vestía sus mejores galas nocturnas recibiendo al conjunto lírico de Bracalle y a la prima donna Iris.
>
> Martinez 1988: 11–12

The problems outlined in the previous chapter evolve from particular historical political ecological processes. As the urbanization process is predicated upon the mastering and engineering of nature's water, the ecological conquest of water is an integral part of the expansion and growth of the city. At the same time, the capital required to build and expand the urban landscape is itself, at least in the case of Guayaquil, generated through the political ecological transformation of the city's hinterland. In this and the following chapters, we shall explore the historical dynamics of the urbanization process through the lens of this double ecological conquest. The city's growth created the need for water systems, which stretched further and further from the city in order to tap additional water resources. Foreign capital had to be generated to finance the imported technology of these projects. This necessitated a sound export-based economy, initially driven by cocoa (until the early twentieth century), bananas (from the mid-1950s to the early 1970s), and oil (from 1973 onwards). The urban process was consequently embedded in a double ecological conquest: ever greater flows of water became urbanized, while the city's hinterland was socially and ecologically transformed. The latter conquest, in turn, plugged the Ecuadorean economy into the international division of labour. Guayaquil was the arena and medium through which those circuits of transformed nature and money were organized.

The contemporary social struggle around water is evidently the result of the deeply exclusive and marginalizing ways in which political, economic, and ecological power have been worked out. The current water system and water politics exemplify the wider socio-economic and political processes that characterized Guayaquil's urbanization process.

4.1 The making of the Ecuadorean bourgeoisie and the first urbanization of water

4.1.1 The origins of the commodified watering of the city

Until the mid-nineteenth century, Guayaquil was just a large port village on Ecuador's Pacific coast, surviving in the shadow of the political and former colonial centre of Quito and the economically dominant Sierra (Andean highland) *hacienderos*. In 1780, Quito had a population of 28,500 compared to 6,600 in Guayaquil, and by the mid-nineteenth century these figures had risen to 36,000 and 25,000 respectively. Until 1700, the potable water supply for the approximately 5,000 residents of Guayaquil came from wells dug at the foot of the Cerro Santa Ana. Later, when water demand outstripped the supply of the wells, the 'local' water had to be complemented by commercialized water transported from the Daule River. The water was captured approximately 25 km upstream from the city, because of downstream saline waters (Estrada 1974), and is the current site of the main treatment station of 'La Toma' (see Fig. 3.4). Professional indigenous 'aguateros' or water vendors[1] transported the water by rafts and mules carried the barrelled water around town (Estrada 1972: 50). Speculative water politics were quite common, ranging from the formation of cartels to increase the water price to the selling of (more saline) water captured downstream, thereby saving on time and transportation costs. From 1739 onwards, the local authorities organized regular raft journeys to haul water from the river, and by the beginning of the nineteenth century, water vending was a very lucrative business as water haulage and distribution fell increasingly under market rules. While other small Ecuadorean towns still had local wells or reasonably clean local river water, Guayaquil depended more and more on an

[1] The local name for these water vendors was 'mitayos' (Perez Pimentel 1987: 116).

already fairly sophisticated social and material water economy that combined private and public actors.

Although several attempts were made to regulate the 'aguateros' or 'aguadores' by setting price levels, prices remained high. The considerable returns of water businesses attracted new and more powerful economic actors into the water economy. Even the ethnic composition of water vendors began to change. Originally an activity organized by indigenous people, the circulation of water became increasingly organized by and under the control of mestizos (people of mixed ethnic descent) or whites. Indeed, while water haulage and distribution originated as a fairly marginal activity whereby indigenous proto-entrepreneurs could find a business niche that permitted basic survival, these were driven out by mestizos and whites as the revenues, and hence the power of water, increased. By 1890, the monthly wage of a water carrier was about 24 sucres (Pineo 1996: 60) and comparable to that of a skilled worker. The changing ethnic composition of the division of labour in the organization of the production and delivery of urban water was paralleled by a deepening transformation of the divisions in terms of water consumption. The spreading commodification of water resulted in a social stratification of water consumption as this depended increasingly on people's ability to pay. The poor were forced to use (dirty) well water or 'hauled muddy drinking water out of the Guayas river from the very piers that also served as toilets' (Pineo 1996: 102). More wealthy inhabitants could afford to pay for the river water or to use expensive porous stones to purify the water (Pérez Pimentel 1987). Estrada (1972), for example, notes how access to water became socially highly stratified:

> To quench their thirst, the rich drank sangria; the middle class water from the Daule River, and the poor. . . . they drank badly tasting water from the pumps in the old city or from the river . . . stored in a pot or filtered through large stones. (my translation)[2]

This basic and decentralized, but commodified, system of untreated crude urban water supply remained in place until the end of the nineteenth century. A first attempt at constructing a sewerage system was made in 1859 when city-workers dug a 10-mile open channel to carry excrement into the laguna. However, this proved to be totally inadequate and dung heaps in the city grew bigger and bigger (Pineo 1996: 101). A first sewer line was completed in 1892, but this had broken down completely by 1903. However, the variety of efforts to sanitize the city that were emerging in Europe had captured the imagination of the enlightened Guayaquileño elites who ventured to modernize their own environments on the basis of the European experience. Trade links, and the frequent travels of the nascent Ecuadorean bourgeoisie to Europe, fused the modernizing visions and engineering capabilities that had swept through the old continent with the aspirations of the local elite to mobilize this cultural

[2] Para calmar la sed de los ricos tomaban sangria; la clase media agua del Daule; y los pobres. . . . agua 'desabrida al gusto', de los pozos de la ciudad vieja o del Rio . . . decantada en una olla o filtrada a travès de grandes piedras.

capital as emblematic manifestations of their own social and political economic ascent. The first studies to equip the city with a circulating running water network were undertaken, and the contest for control over and domestication of water began to intensify (Manrique 1940).

The first urban water engineering studies were initiated in the second half of the nineteenth century, but they did not attract sufficient interest (or funds). Between 1823 and 1885, for example, several attempts were made to start large water projects in Guayaquil, all of which failed because of a combination of systematic rejection of national (state) support and the absence of local interest and financing. This reflects Guayaquil's lack of financial economic power and political influence at the national level during the early years after independence (1830). In 1880 the local authorities contracted a team of engineers to build the first waterworks, including the construction of reservoirs and a pipe network. However, this attempt at constructing a public water supply failed, as local investors did not show any interest in participating in the planned 'Empresa de Agua Potable' (Potable Water Company). In 1884, at the height of the cocoa boom, a public tender was put out to initiate water and canalization works. The financing of the project (estimated at 716,000 pesos), which was started in 1887, was secured through a loan from the Banco de Crédito Hipotecario after the local authority agreed to buy a plot of presumably worthless inundated and marshy land owned by the bank. This land did not yield any rent at the time, but after 1945 became part of the invaded settlement of Suburbio (Pérez Pimentel 1987: 120–1; Rojas and Villavicencio 1988; Villavicencio 1992). The deal allowed the bank to cash in on the worthless land[3] and to use the generated rent to finance the waterworks.

This first water project was executed by a French company employing a series of mostly foreign (German, Italian, British, French) engineers. The reservoir on the Cerro del Carmen was inaugurated on 6 July 1892, and was filled the following day with water piped from a point on the Agua Clara river 88 km east of the city (see inset A on Fig. 3.4). By New Year's Day 1893, the first distribution network was in place and from 30 January of that year, the first 150 houses could enjoy the luxury of domesticated water. In the subsequent years, the network was gradually extended by the local authority under the management of the 'Junta de Canalizaciòn y Proveedora del Agua Potable de Guayaquil', created by President Eloy Alfaro in 1896 and partially financed through a tax on cocoa exports (Villavicencio, Rojas, and Olaya 1988).

Lavatory import businesses were set up very shortly afterwards, and flourished despite the fact that waste water still had to be collected in buckets or allowed to flow freely over the patio to be absorbed by the ground. Toilets and indoor plumbing became valued symbols of cultural capital and testified to the social status of the residents. The urban poor visited the 'sanitized' houses to

[3] It is exactly on these marginal municipal lands that the post-1930 land invasions would begin.

marvel at the imported and finely decorated porcelain or tiled artefacts for bodily hygiene and cleansing (Pérez-Pimentel 1987: 123; *El Universo* 1979). Both city and body joined the conquest for a sanitized, hygienic, and deodorized being. Status, gender and power became reflected by the odours of the body. As Pineo (1996: 74) put it:

> The wealthy adorned their homes with telephones, indoor plumbing, refrigerators, typewriters, various Westinghouse kitchen appliances and even gas lights and stoves. Special imports included American whiskey such as Old Grand Dad, champagne (at four Sucre a bottle in 1925), fine soaps, or Ever-Sweet underarm deodorant, they could purchase at better stores.

The domestication and commodification of water and the associated stratified and exclusionary water practices placed urban water control and use squarely into the realm of social differentiation and status. This, in turn, brought water appropriation even more to the foreground in marking and consolidating relations of power and class positions. While the white rich would defecate in the sometimes silver bowl of the toilet, comforted by the privacy of their custom-made decorated lavatories, and perfumed men and women would promenade along the waterside boulevard and visit the theatre, the poor continued to use the streets as a public toilet, and the river for essential bodily hygiene. Pineo (1996: 102) recalls the memoirs of a visitor in 1914 who wrote that filth continued to 'accumulate . . . in the houses and patios, or courtyards, especially those of the poorer classes and wastewater from taverns, factories, laundries, and homes slopped all over Guayaquil. People urinated and defecated wherever they found a place as the urge came.'

4.1.2 Domesticating water: a double ecological conquest

The mobilization of the city and the state around a growing preoccupation with the water urbanization process paralleled a changing socio-spatial class situation and a reconfiguration of the state apparatus at the turn of the century. In fact, the successful watering of the city at the start of the twentieth century signalled and reflected rapid changes in the political economic position of the city and its ruling elites. Indeed, after independence, and particularly from 1850 onwards, the early post-colonial society underwent significant socio-spatial changes as Ecuador was gradually transformed into an agro-export economy. The agro-export-based Ecuadorean accumulation model originated with the expansion of world demand (particularly in Europe) for and trade in cocoa around 1860. The growth of cocoa production in the coastal region of the country and the concomitant rise of cocoa exports reshuffled the social composition of Ecuador. The country had hitherto been characterized by an economically and politically dominant class of Sierra *hacienderos* on the one hand and an impoverished, politically excluded, and unfree, sharecropping (*huasipungo*) indigenous peasantry on the other. The most widespread form of

Table 4.1. Population change in Guayaquil, 1500–1990

Year	Total	Source	Year	Total	Source
1537	150	a	1880	36,000	a
1571	320	c	1890	44,792	b
1587	786	b	1895	55,000	b
1605	1,100	d	1896	58,000	a
1620	2,000	c	1905	81,650	b
1678	6,000	c	1910	82,000	h
1693	5,000	a	1919	91,842	b
1734	11,000	a	1920	100,000	a
1764	4,914	h	1930	116,047	f
1793	8,000	c	1935	135,190	f
1805	14,000	a	1944	200,000	f
1814	15,000	a	1950	258,966	e
1820	20,000	a	1962	510,804	e
1857	25,000	a	1974	823,219	e
1875	26,000	b	1982	1,199,344	e
1879/82	31,972	b	1985	1,469,353	g
			1990	1,655,592	e

Data for 1990 combine Guayaquil and Duran.

Sources: (a) Chávez (1944); (b) Hidalgo (1932); (c) Estrada (1972); (d) Hamerly (1973); (e) INEC, Census 1950, 1962, 1974, 1982, 1990; (f) PREDAM (1976); (g) Rojas and Villavicencio (1988: 181–2); (h) Pineo (1996: 2).

labour was forced work by peasantry (3 to 5 days per week) on the land of the Andean *haciendero*, a system that was finally abolished only in the 1970s.

By 1890, however, cocoa accounted for 90% of total exports, and in 1904 Ecuador became the world's leading exporter of cocoa (Aguirre 1984; Chiriboga 1980: 261). The old coastal socio-ecological complex, originally characterized by small-scale subsistence production cultivated by a largely endogenous peasantry, had given way to immense cocoa plantations. Cocoa production needed relatively little capital investment and the local elites could mobilize sufficient capital to set up cocoa plantations and export businesses. A variety of forms of waged work characterized capital–labour relations on these plantations. The forced and rapid formation of a wage-dependent class, combined with a disintegration of the peasantry, fed not only the growing demands for wage labour in the coastal plantations, but also for auxiliary waged functions in the city. Between 1896 and 1909, for example, the population of Guayaquil grew by 2.5% annually (Rojas and Villavicencio 1988: 22) (see Table 4.1).

The rapid accumulation of cocoa capital, and the enrichment of coastal cocoa producers, cocoa exporters, and their financial support centres resulted in the rise of an urban merchant and financial bourgeoisie in Guayaquil, the increasing monetization of everyday life and its social relations, and the affirmation and later consolidation of the economic position of the city (Bock 1988; Chiriboga 1988). The rise of the Guayaquileño metropolis was predicated on the transformation of nature and the integration of a new cocoa-based agricultural ecology in the process of production and rent extraction. The rural and urban social and physical environment were both transformed through this social-ecological conquest, which inserted the coastal region squarely into a worldwide monetary circulation process and produced the city as the nexus for cocoa rent appropriation, accumulation, and uneven distribution. The engine of Guayaquil's growth during the first decades of the twentieth century were the copious cocoa harvests in the coastal region surrounding the city, and its insertion into a global trading system.

In addition to serving as Ecuador's main gate to the Pacific and the world market, Guayaquil became the regional centre for cocoa producers and concentrated the key merchant and financial functions, together with their support activities. The circulation of money as land rent, commercial rent, and financial interest was materialized in the bourgeoisification of the city and the rise of an urban rentier class. By 1909, 90% of cocoa production and 80% of all Ecuador's exports passed through Guayaquil. A local comprador bourgeoisie of landowners, merchants, and financiers, together with the mainly European cocoa importers, shared the rents extracted from the migrating workers who transformed nature's ecology and shaped the cocoa-based, and produced, second nature in the lowlands around Guayaquil. The realization and appropriation of these cocoa rents and their reinvestment in the urban environment defined both the form and function of the city and the social position of its promoters (Allou 1987: 25). The monetization of labour relations (which did not occur in the Sierra) in turn fed migration to the coastal cocoa haciendas and to Guayaquil. In addition, many small and medium-sized peasant farms were squeezed out as the consolidation of cocoa haciendas was accompanied by a massive concentration of land, bringing the socio-ecological transformation of nature's production process under the control of a small coastal oligarchy.

By the end of the nineteenth century, 20 families controlled 70% of the land in the cocoa region (Chiriboga 1988: 64). Table 4.2 summarizes the net worth and economic importance of the 'comprador' bourgeoisie of Guayaquil in 1901. Newly established banks ploughed the cocoa rents produced in the city's socio-ecologically transformed hinterland back into the circulation of money, which was in turn recycled mainly through urban investments. Indeed, in the wake of the bourgeoisification of Guayaquil, the first merchant banks were set up. In 1868, the Banco del Ecuador (later Banco de Guayaquil) was established by Lautaro Aspiazu and Lisimaco Guzmán, both of whom belonged to the agro-export group. In 1872, the Banco de Crédito e Hipotecario was set up, which, among others, issued bonds to finance the engineering works for the Guayaquil Water Company (Agua Potable de Guayaquil) (Bock 1988: 28). This was followed in 1895 by the formation of the Banco Agricola y Comercial,

Table 4.2. Number of businesses and their net worth, Guayaquil, 1901

Type of business	Number	Worth (1,000 sucres)	% of Guayaquil's capital assets
Banks, savings and loans institutions	7	13,618	39
Importers	391	8,831	25
Exporters	30	4,429	13
Financiers	141	2,313	7
Fire insurance companies	14	2,230	6
Companies	13	1,145	3
Factories	34	852	2
Steamboat companies	19	376	1
Retail lumber	11	357	1
TOTAL		34,801	

Source: Pineo (1996: 65).

also by the Banking-Agro-Exporter group (Carrión 1991). The concentration of cocoa rents in the financial sector and the subsequent loans made to the national government gave the local bourgeoisie a strong leverage on the national state (Allou 1987; Moncayo 1974: 13), which also contributed to the waning of the Sierra elite's hegemony. The latter had been negatively affected by the downturn in the textile sector and began to feel the sting of a rising coastal elite, whose political ambitions rose alongside their growing economic and financial muscle. By 1924, for example, 80% of the state's debt was held by Guayaquileño banks (Moncayo 1974: 113). In addition to its growing influence at the national scale, the commercial and financial rentier bourgeoisie also began to control 'hegemonic urban institutions' such as 'La Junta de Beneficiencia', 'la Cámera de Comercio', the clubs, the Masonic societies, the newspapers and, even some schools (Quintero 1980: 85–6).

The emergence of the Guayaquileño metropolis was indeed based on the transformation of nature and the integration of a new cocoa-based agricultural ecology into the process of production and rent extraction. The social and ecological metabolism through which cocoa became urbanized in Guayaquil, and subsequently integrated in global flows of cocoa beans and money fused nature and society together in new ways that would shape and transform the city's urbanization process. At the same time, the map of political power was redrawn in new ways. Despite the rapid growth of the city during the second half of the century (70,000 inhabitants by 1900), which led to its size surpassing that of Quito, political power remained in the hands of the Sierra landowners for most of that early period. Nevertheless, the emerging coastal bourgeoisie increasingly challenged the hegemony of the Serrano landed 'aristocracy'. The consolidation of the position of the rentier comprador bourgeoisie towards the end of the century was paralleled by a marked political geographic and ecological shift:

[The origins of capitalist development] show a distinct regional character, because it started in the humid and tropical plains of the coastal region as a result of its insertion into the world market and the international division of labour of the nineteenth century. In the meantime, the Sierra highlands maintained the inherited colonial social structures without major changes (based on the 'huasipungo' hacienda system). Therefore, it is in the coastal plains that a bourgeoisie is formed. Later a bourgeoisie will emerge in the Andean region as well, but its historical origin differs from that of the cocoa-related export bankers and of the Guayaquileño merchants and industrialists (Guerrero 1980: 11–12). (my translation)[4]

Through these political economic and ecological shifts, the urban commercial and banking bourgeoisie in alliance with coastal landowners and cocoa producers now controlled the city and the countryside and began to aspire for more political influence at the national level (Conaghan 1988; Conaghan and Espinal 1990). The coastal political elite increasingly challenged the hegemony of the Serrano landed 'aristocracy' (Guerrero 1980). Guayaquil's lead in the 'Revolución Liberal' of 1895 that propelled Eloy Alfaro, a coastal liberal, to the presidency of the country, would eventually displace the traditional elites from their hegemonic control over the central state institutions. They created a new regional configuration, this time dominated by the lowlands, through the transfer of power to the merchant bourgeoisie:

The commercial and financial bourgeoisie of Guayaquil achieved control over the whole national economy. Then, leading a broad and heterogeneous alliance of diverse social groups, they launched on to the conquest of political power. (Muñoz Vicuña 1987, also cited in Bock 1988: 27) (my translation)[5]

The 'Liberal Revolution' did not so much bring about new forms of political action or intervention as shift the power geometry from Quito to Guayaquil, while the booming cocoa economy meant that the new elites could marshal a much greater volume of capital (Hurtado 1985; Rodríguez 1984). The rapid urban accumulation of cocoa rents, growing residential and functional segregation, and the consolidation of a new economic, cultural, and political elite transformed the state at both the local and national level. The transformation and appropriation of nature by means of a reconfiguration of social relations

[4] [Los origines del desarollo capitalista] presentan un carácter regional marcado, pues se ubican en la planicie húmeda y tropical de la Costa, como consecuencia de su inserción en el mercado y división de trabajo mundiales en el siglo XIX; mientras el callejón interandino, la Sierra, conserva sin modificaciones mayores las estructuras sociales de herencia colonial (basadas en la forma de hacienda huasipunguera). Es, por consiguiente, en la Costa donde se constituye por la primera vez una burguesía. En forma tardia brota tambien una clase burguesa en las ciudades andinas, pero su origen histórico difiere de aquel de los banqueros exportadores del cocoa y de los comerciantes e industriales guayaquileños.

[5] La burguesiá comercial y bancaria de Guayaquil logró un control sobre el conjunto de la economia nacional. Entonces, liderando una amplia y heterogénea alianza de diversos grupos sociales, se lanzó a la conquista del poder politicó.

(capital–labour relation rather than the use of forced labour) and the accompanying expansion of the agricultural frontier in the hitherto only sparsely populated and utilized humid coastal region redefined the geography of Ecuador and shaped political ecological conditions for the years to come. The Guayaquileño elites, now in control of both the local and the national state and closely allied with international merchant (cocoa) capital, generated an uncontested urban growth coalition geared towards securing and promoting their personal, and hence the city's, interests. Indeed, the expansion and restructuring of the city acted as a catalyst for rent and profit appropriation and for the launch of a struggle by the new elites to command the heights of the national state:

> The Liberal Revolution . . . began to use part of its income for financing infrastructure, public buildings, etc., and the construction industry accelerated relatively the absorption of the labour force . . . It is not industrialization that resulted in the concentration of population in this growth pole [Guayaquil] . . . but fundamentally the development of commercial, construction, and service activities. (del Campo 1980: 370) (my translation)[6]

From this time onwards, the urbanization of Guayaquil becomes the history of the conflicts and interests of the urban ruling elites, the alliances that are forged and broken down, and the struggle of this group to control the national state. The fortunes of the urban elite are an integral part of the relationship between the further ecological conquest and nature's transformation in the countryside on one hand, and the position of the urbanized elites in the increasingly globally organized circulation of money and capital on the other. What Michael Johns suggests for elite formation in Buenos Aires also holds for Guayaquil:

> [Its] ruling elite [is] evidently provincial, though eminently cosmopolitan. This rentier class was sustained by agrarian and urban rents, the profit of merchant's capital, the interest earned on credit and earnings on speculative ventures. [They] were dependent on Western Europe for capital, markets and industrial commodities, as well as style, taste and ideas. The rulers . . . lavished much of their extraordinary wealth on a cosmopolitan style of consumption designed to imitate, if not outdo, the elite of Paris and London. (Johns 1993: 75)

Or, indeed, those of Guayaquil as suggested by Martinez's quote at the beginning of this chapter (see also Wolf 1892; Enock 1914).

The new urban bourgeoisie forged a growth coalition, which began the process of rapid urban development through the accumulation, investment, and consumption of the rents and riches from cocoa production (Carrión 1986; del Campo 1980: 370). The form, ideology, and aesthetics of this new

[6] La Revoluciòn Liberal . . . comienza a utilizar parte de sus ingresos en infraestructura, edificios publicos, etc., y el sector de la construcciòn dinamiza relativamente la absorbciòn de mano de obra . . . De forma que no es la industrializaciòn la que concentra poblaciòn en este polo de crecimiento . . . sino el desarrollo de actividades de comercio, construcciòn y servicios, fundamentalmente.

urbanization process were characterized by a dismissal of past forms and materials, the introduction of 'progressive' European ideas and the secularization of urban life, disseminated by the many immigrants and the new Ecuadorean bourgeoisie who had picked up these modernist ideas on their many travels to the old continent.[7] Immigrants, mainly from Italy, Spain, Germany, and Britain, made up 10% of the population by 1910. In addition to investments in urban sanitation projects, roads, public buildings, parks, and monuments, the development of the Quito–Guayaquil and local railroad lines were major achievements of the city's ruling growth coalition (Rodríguez 1987). This reconfiguration of class relations resulted in a series of contradictions, which were contained, among other ways, by public sector organization of urban services and infrastructure. The beautification of the city centre became a prime objective for the bourgeoisie, as the urban proletarianization process began to threaten both their aesthetic and physical sanitary well-being. The Eurocentric upper classes transposed the principles they found on their regular trips to Paris and London to their own urban environments:

> The public works in the city responded to two different demands: one that prioritized the sanitation of the city (potable water, canalization, landfill, etc. . . .); and the other that required the embellishment of the port city for the celebration of the Centenary of Independence. (Rojas and Villavicencio 1988: 85) (my translation)[8]

Moreover, the increasing demand for wage workers in the urban economy created the need for low reproduction costs and relative social peace. The first working-class organizations arose alongside the development of the bourgeoisie. The carpenters' society, for example, was founded in 1896 and went on strike the same year to demand a 9-hour working day. In 1907, a major strike by the railroad workers paralysed the city (Allou 1987: 28). The provision of low cost collective consumption infrastructure (housing, transportation, schooling, urban services) fitted the double purpose of suppressing both the wage bill and social stress, while turning Guayaquil into an exemplary city in terms of aesthetic and sanitary ecological conditions. In addition, the collective nature of such necessary infrastructure and the often externalized returns of such services pointed to the state as the preferred body to initiate, organize, and control such ventures. As the majority of the people were excluded from the political decision-making process anyway, the state was able to take a class-biased position, which permitted and perpetuated bourgeois control over the state's institutions. Urban engineering and construction works were initiated within this framework, with water production and distribution, combined with

[7] The relationship between the rise of the Guayaquileño bourgeoisie and the changing city architecture and accompanying aesthetic views is detailed in Sophie Bock's excellent study (1988). (See also Godard 1988.) For a similar study of Buenos Aires, see Johns (1992; 1993).

[8] Las obras publicas de la ciudad responden a dos demandas diferenciadas: la una, que priorizaba el saneamiento de la ciudad (agua potable, canalizaciòn, relleno, etc.); y la otra, que exigia el embellecimiento y ornato del puerto con miras a la celebraciòn del Centenario de la Independencia.

a late attention to sewerage, as key priorities. Around the turn of the century, the urbanization and domestication of water in Guayaquil kept pace with the accelerating urbanization of capital. The very history of the mastering and domestication of water to supply the urban realm narrates this story of changing urban social positions and the emergence of new urban conflicts.

As the city expanded (see Table 4.1), however, sanitary and physical conditions deteriorated. Eloy Alfaro, uncontested leader of the liberal revolution and President of Ecuador, decided to create the 'Junta de Canalización de Guayaquil'. In 1900, he declared the urban water project and other sanitary infrastructure as works of national importance to be financed largely by the national state on the basis of taxes levied on the cocoa trade. In 1905, another special cocoa-export tax was levied to finance the sanitation works as well as other major infrastructure projects such as the riverside promenade (Malécon), the central park, the transportation system, public offices, and buildings. Between that time and the 1920s, the water system was gradually extended, slowly following the pattern of urbanization, which nevertheless increasingly began to outpace the expansion of the system. Indeed, the basic system put in place at the start of the century (supplying on average 2,500 cubic metres per day) would remain the city's main source of water (of questionable quality) until 1928 (Acevedo 1938; Manrique 1940). Nevertheless, the domestication of water in Guayaquil had outpaced the sanitation of many European towns of a similar size, some of which did not have the equivalent facilities until well into the twentieth century (Goubert 1989).

From the early years of the water system, charges were differentiated on the basis of activities within a building and estimated water consumption. Water use was not metered, but instead the diameter of the water pipe (1 unit = 1/8″) served as the main accounting unit. In general, commercial activities were charged more than private houses or offices, while industrial activities had to pay the most (indicating the particular form of elite coalition who accumulated money through rent and interest extraction rather than through production). Indeed, commercial or related activities paid considerably less than industrial activities even if the former used huge quantities of water. A form of progressive pricing policy was introduced as people in rented accommodation paid less for their water than homeowners. Public institutions (offices, hospitals, schools, etc.) received water free of charge.

However, while for the upper classes Guayaquil prospered during this period and revelled in its new leadership role (Martinez 1988), the seeds of the subsequent disintegration soon germinated. In fact, from the second decade of the twentieth century onwards, the moment when population growth outstripped the expansion of the water network and the hegemonic power of the local elites began to show its first internal fractures, private water vending again became an accepted part of daily life. From this point onwards, the excluded will constantly suffer the difficult daily quest for water. However, this also marks the beginning of the private water sellers buying water from the public utility for distribution and sale in areas that lacked basic water infrastructure. The officially charged price to water vendors was very low, permitting them a quite considerable extraction of monopoly rents.

Faced with a growing gap between supply and demand, the municipality proposed and contracted work to increase the water production and water conduction capacity from the Agua Clara source to 13,000 m³/day, sufficient to provide a city of 150,000 inhabitants with an average of 100 litres per person per day (*Gaceta Municipal*, November 1910: 167; 21 February 1911). However, a local inspection team visiting the catchment site in 1916 reported a physical water loss of 50%, and the pipeline to Guayaquil showed serious problems and damage (*Gaceta Municipal*, 20 March 1916). Both the municipality and the Junta de Canalización y Proveedora de Agua acted as contractors for the French consultant engineer E. Coignet. This dual contracting suggests that both institutions were formally independent bodies. The interest on and repayment of the loans raised to cover project costs were paid by a combination of local land taxes, the sale of water, and taxes on imports and cocoa exports. This mechanism again demonstrates that the financing of the water urbanization projects was by no means self-supporting and largely depended on the level and vitality of the import/export sector and the continuation of an accumulation process based on cocoa metabolism, and on the extract of land or commercial rent. It suggests that investment costs could not be recuperated from the sale of water alone, and therefore indicates a substantial gap between the cost of water and the price charged to the consumer. Indeed, from the beginning (and until this very day), the expansion of the water system has been dependent on external financial sources (and, therefore, closely related to the capacity of the Ecuadorean economy to generate foreign currency via export promotion), and operational costs have depended on a combination of loans and subsidies (see Chapter 6). Of course, this highly subsidized water supply system suppressed the cost to the consumer, thereby satisfying the demands of both poor and rich for low-cost access to potable water.

Despite these efforts, Pineo (1996: 103) reports that Guayaquil in 1921 received only a fifth of its required water supplies. As one observer put it: 'Guayaquil, surrounded by water, nevertheless finds itself short' (cited in Pineo 1996: 103). In any case, the city, at least partially washed with water streaming through its veins, cleansing its burghers, and portraying an image of health, beauty, and prosperity, faced its first major water crisis towards the end of the first decade of this century. The commodification and urbanization of water, however partial it may have been, nevertheless gave water an increasingly prominent place in maintaining the social fabric of urban life. The city had grown on the basis of an ecological conquest and appropriation of the rents from agricultural produce. This was paralleled by the harnessing and urbanization of water, thereby inserting the circulation of water squarely into the circulation of money and its associated power relations and class differentiations. Yet while the city could no longer survive without being perpetually washed with an incessant flow of water, the expansion of centralized water provision began to slow down at a time when the city itself expanded.

4.1.2 Moving the water frontier and the emergence of exclusionary water practices in an age of reformation

It soon became evident that the current catchment would not be sufficient to guarantee a sufficient supply of water to match the pace of the water urbanization process. In 1924, for example, the city received water for only two to three hours a day, usually between 6.30 a.m. and 9.30 a.m. (Manrique 1940). Nevertheless, the Agua Clara river remained the city's main source of water (of questionable quality) until 1928 (Acevedo 1938). If the domestication of water were to continue to follow the expansion of the urban frontier, the water frontier needed to be pushed outward in search of new exploitable water reserves. The growth and expansion of the city could only be sustained by pushing First Nature's frontier further outwards and by incorporating ever larger parts of nature's geography into the circulation of money and profit upon which the city's continuing prominence depended so crucially.

As early as 1916, the London-based J. G. White & Co. presented a plan for the provision of water to Guayaquil from the Eastern Andes cordillera (about 88 km away from the city centre), as well as a sewerage system. This proposal would have increased the capacity of the existing system to 20,000 m^3/day for a population of 117,000 inhabitants by 1926. At that time, the system would need to be upgraded to produce 30,000 m^3/day, which should be sufficient until 1936, after which the capacity would need to be increased to 40,000 m^3/day. On 5 January 1919, the national government contracted the J. G. White company to implement the proposed water supply system, and a few months later the local authorities took over the contract from the national government. In 1923, the 'Junta Especial de Saniamiento' was created to execute public works of canalization, potable water provision, sewerage, and road works. The Junta was financed by local taxes, which were nationally collected and then devolved to the local authority. The White company continued with its works while the government allowed the municipality to contract (foreign) loans to finance the project.

As a result of this, a new production and conduction system, known as 'La Lolita', was inaugurated in 1928 with a capacity of 20,000 m^3/day destined for Guayaquil and the surrounding region (see inset A on Fig. 3.4). From then until about 1932, the city enjoyed a 24-hour supply of water, a unique achievement in the context of Latin America and something the city would never be able to accomplish again. In 1933, when the city had 117,000 inhabitants and almost 8,000 dwellings, the average daily supply of water was 170 litres per person per day. From that time onwards, water was charged on the basis of either metered water consumption or on the basis of cadastral evaluation of the property (Ordenanza sobre Agua Potable 1933). Not surprisingly, this was also the period in which yellow fever virtually disappeared from the urban scene. The early efforts of the Rockefeller Foundation in 1919 had been instrumental in achieving this (González 1988; Pineo 1990).[9]

However, this successful watering of the city was short lived. Very soon thereafter, the water urbanization process slowed down dramatically, as political power relationships began to shift in decisive new ways, particularly after the crumbling of the cocoa economy. Indeed, by the end of the 1930s the highly successful and hegemonic bourgeois growth coalition that had launched Guayaquil on the path of dependent modernization (while the Sierra had relatively stagnated) had dwindled. The high rents from the cocoa economy had put the *cacaoteros* in a position that did not tolerate challenge, both locally and nationally, while the continuous success of the rentier export economy prevented internal differentiation, as there was no need to explore new ways of accumulating money. Nevertheless, the cocoa boom began to collapse at the beginning of the 1920s, and would eventually plunge Ecuadorean society into a great economic and political crisis. This collapse also produced the first cracks in the elite alliance of cocoa producers, merchants, and financiers. Between 1917 and 1926, for example, cocoa output fell by 45% from 1,008,000 quintals to just 447,000 (Bock 1988: 60; see also Crawford de Roberts 1980). The socio-ecological opening up of Africa for cocoa production for the world market, the phyto-sanitary problems resulting from monocultural practices, and the dwindling European demand for cocoa during the First World War had a negative impact on prices, productivity, and production. Between January and September 1914 alone, cocoa prices on the world market fell by 50% (Martinez 1988: 19). The breakdown of the global economy after the war further eroded prices, with the cocoa price on the New York commodities exchange plummeting from 25.75 dollar cents per pound to 5.75 dollar cents between March 1920 and June 1921. By 1921, the amount of cocoa export revenues had fallen by 21% compared with those in 1917. As prices fell, foreign currency became scarce and capital for domestic investment began to dry up. The subsequent dollar shortage hiked up the price of the dollar and cocoa exporters were able to recapture some of the lost ground as a result of the increasing exchange rate (from 2.11 sucres to the US dollar in 1920 to an average of 4.2 in 1922 and 6.2 in 1923). This, in turn, put the import sector into serious difficulties and resulted in monetary chaos (Marchán 1988; 1991). In 1923, the price for cocoa on the world market hit rock bottom and was only half of

[9] Indeed, the 'annihilation of space by time' had brought the US much nearer to the lethal diseases of the Latin American subcontinent. Merchant ships from Ecuador and other places not only brought agricultural produce to the US market, but diseases and their vectors as well. Yellow fever appeared in North America. Eradicating the disease implied sanitizing the coastal areas and, in particular, the port cities. The Rockefeller foundation initiated and completed an eradication programme in Guayaquil with great success.

its value between 1900 and 1913 (Pineo 1996). Moreover, the Ecuadorean cocoa harvest declined significantly between 1923 and 1924 and not only as a result of falling world prices. The monocultural cocoa plantations were badly maintained as copious harvests, high cocoa prices, and abundant cheap labour militated against the need for sophisticated phyto-sanitary intervention, and were thus highly susceptible to the spread of diseases. Two major blights hit the cocoa fields: an epidemic of Monilia fungus in 1914, and witchbroom disease in 1922. In no time at all, previously healthy trees were turned into dry and dead sticks, and in some places, almost all the trees died within a year (Pineo 1996: 145). The combined result was a massive reduction in export earnings for Ecuador and this 'rapidly translated into decreases in public revenues for public finances' (Rodríguez 1984: 175), which hit Guayaquil and urban investments particularly hard. Ecuador had borrowed heavily against the future during the years of the cocoa boom, betting on continued expansion of production, growing export sales and high prices (Marchán 1988).

The stuttering urbanization of water during this period was related to and expressive of fluctuations on the New York commodities exchange market, the vagaries of the international monetary system, and changing socio-ecological processes (disease-ridden cocoa plantations) in the coastal region around Guayaquil. Disease, commodity prices, the expansion of Africa's cocoa production, and the contradictions of public and international finance fused together in a myriad of intricate ways to produce an urban crisis that would quickly sharpen social tensions and accentuate the already problematic socio-ecological conditions in the city. The rapid downturn of the cocoa economy—which at its height had prevented other capital factions from rising to the challenge of the *cacaoteros* in any serious way—left a vacuum which could not easily be filled. Consequently, the hegemony of the bourgeoisie was increasingly challenged from within as the crumbling of the cocoa boom increased tensions between the various local elite rentier factions. For example, in 1917–18, a few members of the import bourgeoisie founded the Banco la Previsora, which would focus its activities on real estate credit and became part of a nascent alliance of real estate developers and the construction industry (Guerrero 1980: 197). Until that time, the leading bank had been the Banco Comercial y Agrícola, which had close ties to the 'cocoa' bourgeoisie (Estrada 1982; Dillon 1985; Chiriboga 1988).

In light of diminishing accumulation opportunities in the cocoa business, some factions of capital turned to other sectors of the economy. Recycling capital through the built environment became one of the preferred niches for profit making. This was a first challenge to the hegemony of the comprador bourgeoisie and an attempt to secure the survival of the ruling elites and the position of the city. The limited internal buying capacity of the masses, however, ruled out the possibility of an upturn on the basis of an emergent national bourgeoisie. This first crack in the urban hegemony was soon followed by other strategies that were deployed by elite factions in order to maintain their eroding position. Urban landowners, for example, subdivided inner-city houses and rented them to the large numbers of immigrants ('tugurización') in a desperate attempt to maintain the accumulation process (Rojas and Villavicencio 1988) by shifting capital into other circuits of circulation (Harvey 1985). Immigration accelerated rapidly as the disintegrating rural cocoa economy shed large numbers of workers who flocked to the city in search of survival and jobs. In addition, internal cohesion was further undermined when the financial elites de-linked the sucre from the gold standard in 1914, resulting in inflation and devaluation of the local currency and negatively affecting the position of the importers (Marchán 1991). The cocoa producers, the weakest link in the alliance anyway, were almost completely destroyed by the collapse of the market. In short, the diverse factions that comprised the urban bourgeoisie began to pursue their own specific interests (housing, land development, or financial speculation) in a desperate attempt to displace immanent devaluation pressures onto other groups and to safeguard their own position. The earlier growth coalition broke down and became rife with internal tensions and conflict. In addition, the Sierra landowners, together with the textile industrialists as the Sierra's main manufacturing sector, seized the opportunity and began to challenge Guayaquileño control over the central state apparatus (Cueva 1974; 1991; Quintero 1980).

The hegemony of the Guayaquileño bourgeoisie was also challenged by the emergence of a local petty bourgeoisie and the growing militancy of organized labour. As early as 1905, the Confederación Obrera Provincial de Guayas had been formed as a regional umbrella labour organization, and the first exclusively wage workers-based union was established in the city (Sociedad Cosmopolita de Cacahueros Tomás Briones) in 1908. The latter declared a strike in that year and again in 1916. After years of economic decline and growing austerity, a citywide general strike—the first major attempt at organized labour struggle—was launched in November 1922, and paralysed the city completely for days. The waterworks were closed down and the city virtually died of thirst. On the orders of the state, the strike was finally brutally and bloodily repressed by the army (Robelino 1976; Martinez 1988), ending with several hundred butchered corpses floating down the Guayas River (Muños Vicuña 1978; Pineo 1988).

With the end of the cocoa boom and the intense social struggle and turmoil associated with the ensuing socio-economic disintegration, a new regional shift in political power started to emerge:

> From now onwards, Guayaquil gave ground as the principal urban centre of the country. Its promoters, now in a process of decline and decadence, did not succeed in overcoming the challenge posed by the cocoa crisis, and in establishing its domination on a new basis. They gradually disintegrated as the dominant class. In 1925, 'progressive' members of the military took power and, in 1933, a new alliance of

landowners gave national power to the populist leader Velasco Ibarro. (Allou 1987: 29) (my translation)[10]

Indeed, the 'Revolución Juliana' of 1925 effectively ended the Guayaquileño elites' control over the state apparatus. The political centre of gravity shifted increasingly (but not exclusively) in favour of Quito and the Sierra highlands as the Sierra landowners reclaimed some of their weakened influence (Godard 1987: 113; Marchán 1991: 48). Since that time, Ecuador has never again experienced hegemonic elite control comparable to the one existing during the era of the cocoa boom. As Maiguashca and North (1991) pointed out, the socio-spatial diversification of both agricultural and manufacturing production contributed to ongoing factional struggles between Costa and Sierra as well as between their internally divided ruling classes:

The different regional and sectoral fractions of the dominant classes succeeded in forming a temporary alliance, but this did not generate the establishment of political and social hegemony by the Sierra landowning class in the medium and long term. To begin with, they were incapable of subordinating, or leading, the economic elites of the other regions. These regional and other struggles within the dominant class were reflected in the political instability of the 1930s and 1940s. (Maiguashca and North 1991: 109) (my translation)[11]

The subsequent period was indeed characterized by political instability. No fewer than seventeen governments succeeded each other during the 1930s, exemplifying the continuous battle between Sierra and Costa and their respective class factions (Deler 1981; Hurtado 1985; Cueva 1991). This factional struggle and instability further eroded cohesion in civil society and, not surprisingly, resulted in growing calls for federalism and regional independence. Particularly in Guayaquil, the weakened bourgeoisie considered regional independence and the construction of a new regionalized state to be a means for restoring some of its lost hegemonic position over the national state (Quintero and Silva 1991). In 1933, Velasquez Ibarra won the elections (something he would repeat four more times), announcing the beginning of Ecuador's subsequent populist tradition[12] (Egas 1992; Menendez-Carrión 1986). In fact, the rise and consolidation of populism as a key political movement—interspersed with military juntas backed by a rivalling diversity of upper-class support—

[10] De hecho, en ese entonces, Guayaquil cedía terreno como principal foco urbano del país. Sus promotores, en proceso de decadencia, no lograran superar el desafío prensentado por la crisis cacotera, asentado su dominación sobre nuevas bases, y paulatinamente se desmorornaron como clase dominante. En 1925 militares 'progressistas' se hicieron con el poder y en 1933 una nueva alianza de terratenientes llevó al líder populista Velasco Ibarra el cargo supremo.

[11] Las diferentes fracciones regionales y sectoriales de las clases dominantes lograron formar una alianza momentánea; pero esto no generó el establecimiento de la hegemonía politica y social de los terratenientes serranos a mediano y largo plazo. Para empezar, fueron encapaces de subordinar, o liderar, a las élites económicas de las otras regiones. Estas luchas regionales y de otro typo, occuridas dentro de la clase dominante, se reflejaron en la inestabilidad política de los años treinta y cuarenta.

[12] This tradition became particularly deeply entrenched in Guayaquil and found a strong popular basis in the invasion settlements (Quintero 1991).

exemplified the absence of a hegemonic elite bloc in power and suggests how a diverse set of class and power interests fused around a populist ideology as the means to secure a minimum of social coherence and stability. Combined with a more regionalist outlook from at least some factions of the bourgeoisie, whose interests were often deeply 'localized' and provincial, the subsequent geopolitical dynamics led to more complex and tenuous relationships between the national and local state (Maiguashca 1991; see also Quintero 1991).

Clearly, declining cocoa rents played a decisive role in the political weakening of the local elites. Combined with the shifting socio-spatial class relationships, which broke the alliance between the national and local state, the weakened Guayaquileño elite found itself on a more confrontational course with the national state. Moreover, during the period of liberal hegemony, taxes, mainly from coastal exports, were going to the national state and were channelled back to Guayaquil by the coastal elites that controlled the state. However, as their control over the national state weakened, nationally appropriated cocoa rents earmarked to finance sanitary and beautification works in Guayaquil were increasingly diverted to other purposes and places.

The loss of power for Guayaquil's ruling classes was paralleled by a slowdown of the water urbanization process. The capital earmarked for extending and upgrading the system was never made available, despite desperate appeals by the local government to the national state and the many promises made by the state to assist Guayaquil's investment programmes (*Revista Municipal* 1936: 52). A 1943 memorandum of the cantonal Council of Guayaquil to the National Congress details the history of appropriation and allocation of cocoa rents (*Revista Municipal* 1943). In 1923, the National Accounting Tribunal (Tribunal de Cuentas) certified that of a total of 32 million sucres of cocoa rents collected between 1899 and 1923 and destined to finance sanitation works in Guayaquil, only 4 million were actually spent. In 1923 a new provincial 'Junta Especial de Saneamiento' was established and undertook to sanitize the city further. The National Congress budgeted an annual amount of 3 million sucres for that purpose. In the subsequent years, 6 million sucres was spent on improving the water supply system. In 1926, J. G. White & Co. delivered plans for road and sewerage works to the Junta de Saneamiento. In 1927, however, the government replaced the provincial 'Junta de Saneamiento' with a nationally appointed 'Jefe de Fiscalización' to oversee the works. The contract with the J. G. White & Co. was taken over by the State and eventually rescinded, leaving the sanitation works unfinished, and showing how the geopolitics of the country change to the detriment of Guayaquil. With the disappearance of the 'Junta de Saneamiento', the city lost the power to control its own economic resources and the capacity to invest these in local infrastructure. The ten million sucres available in the national 'Caja Fiscal' for urban sanitation projects in Guayaquil was spent on other projects by the reformist dictatorial and revolutionary juntas of that time. After the liquidation of the J. G. White & Co. contract, a satellite within the Ministry of Public Works was established under the heading of 'Saneamiento de Guayaquil', which controlled an annual budget of 600,000 to 800,000 sucres. Until 1943, the money was spent mostly on public works throughout the country, with precious little going to improve water and sewerage infrastructure in Guayaquil. In 1929, the National Assembly authorized the Executive Power to borrow the equivalent of US$2 million for the implementation of sanitary works in Guayaquil, although it was not until fourteen years later, in 1943, that the Cantonal Council demanded the execution of the decree. In sum, between 1899 and 1943, the national state received a total of 86 million sucres from Guayaquil's cocoa rents, while only 16 million was actually spent on the waterworks, mainly in the period before 1923. The total national debt to the city was estimated at 76 million sucres (*Revista Municipal* 1948). After 1923, the moment at which political power at the national level was recaptured by highland elites, few if any of the commitments made by the national state with respect to the sanitation of Guayaquil were actually honoured. This struggle over the allocation of and control over investment funds generated through cocoa rents encapsulates the changing spatial and social class composition of the local and national elites and the reconfiguration of socio-spatial power geometries.

4.3 Water, war, and thirst in the 1940s: the selective urbanization of water

In the meantime, the city had experienced rapid growth, with the population increasing by 182% between 1925 and 1950. This led to the development of the first invasion settlements, mainly in Suburbio, whilst the development of the infrastructure stagnated. As Godard (1987: 113) concludes his discussion of Guayaquil:

> It seems that the migratory movements to Guayaquil are inversely related to the realization of infrastructure works, among others because of the economic difficulties the country encountered from 1925 until 1950. (my translation)[13]

The collapse of cocoa-based rural-class relations affected both urban and rural conditions. In the coastal countryside, the large cocoa plantations diminished significantly resulting in a partial exodus of the displaced rural proletariat, which fed rapid urban immigration and constituted the kernel of a growing urban underclass. Agricultural production on the Costa became more diversified, with rice, coffee, bananas, and sugar being cultivated on small and medium-sized farms formed out of subdivided plantations. Migrating peasants could occupy uncultivated plots of land that previously had been owned

[13] Parece que los movimientos migratorios hacia Guayaquil son inversamente proporcionales a la realización de infraestructuras, entre otras razones, por las dificultades econónomicas que el país atraviesa de 1925 a 1950.

by the 'terratenientes' (large landowners). The decline of the cocoa-based coastal liberal hegemony opened a window of opportunity for the relative prosperity of smallholders, who developed a diversified agricultural ecological system, which was nevertheless embedded in a money-based market economy. This period proved to be rather positive for peasants and smallholdings and altered the political ecology of the countryside, while the city began to stagnate economically as its parasitic unity with agricultural export production weakened.

Charles Nurse (1989: 101) summarized this socio-ecological urban–rural transformation process as follows:

> With the collapse of the cocoa boom came the abandonment of many of the spectacular public works projects undertaken in the years of plenty. Also abandoned were many of the plantations themselves, with the owners acquiescing in their takeover by the former plantation workers at low rents or rent-free.

Needless to say, the urban water flows were to become thinner, more unreliable, and eventually grossly insufficient as the circulation of money not only dwindled, but was increasingly diverted to other purposes and recycled through other political configurations (the state) and actors. From 1932 onwards, water supply became gradually more problematic, falling from an average of 20 hours a day in that year to an average of 14 hours in 1938. Water needed to be stored in receptacles of all kinds in order to ensure a 24-hour supply of water (*Revista Municipal*, 1 February 1938: 39–41). Despite several attempts, the two planned expansions to 'La Lolita' were never implemented, while suggestions to construct a second treatment plant on the Daule River (the site of the current treatment plants) also failed to materialize (*Revista Municipal* 1938: 39–41). In 1936, a technical commission reported that the system was in need of urgent technical repairs and that it had suffered damage and water losses as a result of illegal connections and spillage (*Revista Municipal*, July and August 1936: 29–30, 52). In 1939, the *Revista Municipal* noted that the quantity of water produced remained at the 1928 level, although the number of urban residents had almost doubled to about 200,000 inhabitants. The finances for extending the system would never come forward despite the desperate appeals of the local government to the state for money, and the plant of 'La Lolita', with its average daily production capacity of 21,000 m^3, remained the sole source for the city's potable water (*Revista Municipal* 1936: 52). Moreover, the system also supplied water to the towns and villages adjacent to the main conduction pipe from 'La Lolita' to the city centre, so that by 1944, not more than 10,000 m^3 per day actually reached the city for further distribution (Olaya and Villavicencio 1990). This single water-sourcing system remained in place until 1951, and average daily per capita production dropped from about 180 litres in 1930 to less than 100 litres in 1945.

The city's growth was characterized by inner-city densification ('tugurización') and the emergence of the first land invasions, mainly in the tidally inundated mangrove forests of the Guayas estuary (Godard 1988). Ironically, this was municipally owned land that had been bought from a local bank in the 1880s in order to secure a loan for the financing of Guayaquil's first waterworks (see above). The slowdown in infrastructure provision also resulted in the reappearance of slums and a growing number of private water vendors, accompanied by monopoly and exclusionary water practices. The intermittent flow of water, and the difficulties associated with accessing it, increased the danger of contamination, and bacteriological pollution became again a serious threat. In 1938, for example, new cases of yellow fever were reported (*Revista Municipal* 1938: 39–41).

During the Second World War, conditions deteriorated even further. By 1943, the potable water service in the city had diminished to 11 hours a day, falling further to 6–7 hours in 1946. At that time, 'La Lolita' produced 20,000 to 30,000 m³ of potable water each day, with crude water supplied by the Agua Clara, Alvarado, Blanco, and Mayaguán rivers flowing down from the Western Andes cordillero. Two pipes of 11″ and 22″ diameter respectively transported the water to Duran (Eloy Alfaro) on the eastern shore of the Guayas River, just across from Guayaquil city. From there, water flowed through sub-fluvial pipes to two reservoirs on top of the Santa Ana hill. The urban distribution system covered the old, densely populated, centre of the city and with the exception of a few new connections in four more streets in 1946/47, the 1929 distribution network had not been extended. In 1946, only 8,600 dwellings enjoyed the luxury of indoor water, mostly without meters (Buck, Seifert, and Just, Consulting Engineers 1947). While the existence of connections did not in any way suggest the availability of water, there were still more than 110,000 people (55% of the population) who had to make do without any domestic water supply at all (based on a generously estimated average of ten persons per household). An estimated 47.5% of the water was lost through leakage and physical damage in the system (Villavicencio 1992). Those deprived of water lived mainly in the new suburban 'invaded' expansions in the West and Southwest.

The socio-spatially stratified water supply was also locked into the water-engineering system itself. Each of the reservoirs on the Santa Ana hill covered part of the system independently. No bypass system existed, which led to regular interruptions in water distribution each time the reservoir or its mains needed repair or maintenance. The network was organized as a herringbone system (see Fig. 3.4), and water pressure fell rapidly as the water moved away from the reservoir. Close to the feeding station the water supply was fairly regular and of reasonable quality, whereas further down the network, pressure fell and water availability and quality deteriorated quickly until at the margins of the system only a trickle of smelly water remained. The water network corresponded closely to the geography of urban land rents, which were, in turn, reinforced through the spatial inequality organized in and through the technological and engineering structure of the water system. The social geography of the city mirrored the structure of the water network and the level and quality of water delivery. The urban water crisis was spreading rapidly, threatening the very sustainability of the city. Private water vending (particularly in areas deprived of piped water) was common and continued to increase both in numbers and in volume as the water crisis intensified during the subsequent decades.

These exclusionary urban water politics, which shaped the wartime and early post-war geography of the city, were institutionally organized through a municipal public water utility controlled by the 'Departamento de Agua Potable'. The water price was set far below real cost, which resulted in the middle and upper classes receiving highly subsidized water, and the poor being deprived of it entirely. The resulting structural financial deficit jeopardized the maintenance and upgrading of the urban water system and made water-engineering works structurally dependent on financing by means of dwindling agricultural exports.

The political ecology of water production and supply was the outcome of the continuing quest to urbanize water in a context of growing supply problems and increasingly polarized water politics. This process of water urbanization led to an acute water crisis by the end of the forties, after which water problems would never disappear again. On the contrary, exclusionary water politics and water speculation by water vendors increasingly came to characterize urban struggles and became an integral part of the rituals of everyday urban life. In short, from the very beginning, the urbanization of water became subject to intense social struggles as the dialectic of exclusion from and access to commodified urban services unfolded. The perpetual budgetary problems of the state, controlled by an externally oriented, but increasingly internally divided and competing bourgeoisie, resulted in domestic underinvestment combined with a chronic dependence on external funds. The respect that visitors had showed for the city's sanitary conditions during the mid-thirties was now merely a reminder of the past glory of the city elite (Desmarest 1937: 54).

5

The Urban Conquest of Water in Guayaquil, 1945–2000: Bananas, Oil, and the Production of Water Scarcity

'Agua, Drama sin Final' (local newspaper headline).

5.1 Opening up a new waterfront: water and bananas

With the end of the war came a partial reversal of the devastating decline associated with the cocoa collapse, paralleled by a profound reconfiguration of class relations. The pre-war bipartisan political structure (Liberals and Conservatives) was replaced by a myriad of new political parties, expressing the divisions within the ruling elites, the rise of Left political parties as a result of growing proletarianization (Maiguashca 1992: 200–1) and, most importantly, the emergence and spectacular growth of populist movements. New forms of class struggle would emerge out of this maelstrom of change, each expressing itself through a mixture of new and old languages, symbols, and activities. It is not surprising, for example, to hear 'San Lenín' called upon for assistance alongside saints of the more traditional variety (Maiguashca and North 1991: 99–100). The ferment of this rich mix of class relations through which daily life was organized at the time the world was on fire wrought the conditions from which the post-war intensified water conquest would emerge.

Indeed, the turbulent but lean years of the 1940s were followed by the banana bonanza decade of the 1950s. The United States' fruit corporations, their plantations struck by Panama disease, moved their centre of operations from marginal Central American and Caribbean exporters to Ecuador. It was not only a cheap location, but the Panama disease had not yet moved that far south. In addition, President Galo Plaza Lasso used his excellent relationships with the US United Fruit Company to promote banana production in Ecuador (Nurse 1989). The spiralling demand for bananas from the US fruit companies converted the coastal area of the country (La Costa) into large banana plantations with their associated socio-ecological relations (Armstrong and McGee 1985: 114; Larrea-Maldonado 1982: 28–34; see also Schodt 1987). While in 1948, banana export receipts amounted to only US$2.8 million, this figure reached US$21.4 million in 1952 and US$88.9 million in 1960, accounting for 62.2% of Ecuador's total exports (Hurtado 1981: 190; Grijalva 1990; Cortez 1992). By the mid-1950s, the country had become the world's leading banana exporter. This manufactured 'banana bonanza' was organized through a new political economic and ecological transformation. Not only were bananas cultivated on the former cocoa fields or on land that had been abandoned by cocoa producers and occupied by smallholders, the ecological frontier for agricultural export production around Guayaquil was pushed further outward, assisted by an extensive state-funded road expansion programme (León 1992; Trujillo 1992). This transformation radically altered the ecology of the urban–rural complex and incorporated increasingly larger areas into a global money flow.

The general rise in world commodity prices during the early post-war period also stimulated the export economy. Precarious modes of labour utilization and control were gradually abandoned through a series of land and legal reforms. Although production was predominantly organized in smallholdings, the commercialization was concentrated in very few hands, combining a tiny national comprador elite with the US fruit-trading companies (Báez 1985). This banana colonization prompted mass migration to the coastal areas, combined with a rapid urbanization of Guayaquil, whose banana-dependent financial and service economy expanded rapidly (Carrión 1992). Between 1950 and 1974, for example, the city's population grew from 200,000 to more than 820,000, mainly in the newly invaded areas of Guasmo and Suburbio. Once again, the socio-ecological metabolism of Guayaquil's hinterland would fuse with global political dynamics and economic flows.

Banana rents were ploughed back—either directly or indirectly through the state—into the urban realm (Báez 1992). The backbone of Guayaquil's accelerated urbanization process lay within the expanded and reworked ecological conquest of the coastal region, while the latter was itself directly related to the expansion of metropolitan and global agro-business capital. This economic growth improved Ecuador's credit rating and, combined with the efforts of the newly established international financing organizations, foreign capital began to flow into the country once again. However, the agro-export system was organized and controlled by only a handful of families, preventing the rise of a broader social elite. The national state remained fairly weak and internally fractured, while regionalist or populist struggles dominated the political arena, suggesting a symptomatic absence of a more or less hegemonic national elite programme (Conaghan and Espinal 1990). However, in line with the rising prominence of Keynesian forms of state intervention and the limited imitation of European- or US-style 'Fordist' modes of governing, the state would become more directly involved in planning and implementing major projects on the one hand, while the rudiments of an embryonic welfare state system contributed to the emergence of a relatively small but politically significant middle class on the other. The 'peripheral Fordism' (Lipietz 1986; 1987) that emerged reinforced social divisions and formalized divisions between those that participated fully in civil society and those (mainly indigenous people and impoverished mestizos) that became ever more excluded. The fine-grained texture of economic, political, social, and ecological transformations produced a ferment from which the post-war expansion of the urban water frontier to new and hitherto unexploited water reserves would emerge. Banana rents were combined with international loans to finance the rapid urbanization (and peripheral modernization) of the country. It is not surprising, therefore, that this new ecological conquest was combined with a reinvigorated quest for the control over and domestication of nature's water.

On 31 October 1945, the Municipality of Guayaquil signed an agreement with the US-based Pitometer Company, the Buck, Seifert and Just Consulting Engineers (BSJEC), and the Frederick Snare Corporation for the planning, execution, and supervision of the works for the new potabilization plant for the city of Guayaquil. Even the geographical geometry of contracted consultant companies and the sources of financing reflected the shifting geopolitical realities of the post-war world. Whereas European, and mainly British, capital, engineers and companies, were present before 1940, the tide turned unequivocally towards the US after 1945. The national government, under the presidency of Velasco Ibarra, authorized the Municipality to borrow US$4 million from the US Export–Import Bank to finance new water and sanitation projects, and underwrote this loan (Registro Offical, no. 147, 8 Jan. 1946). One year later, a contract with BSJEC was signed to undertake a study 'Informe Sobre el Diseño del Proyecto de Abastecimiento de Agua del Rio Daule para Servir a la Ciudad de Guayaquil, Ecuador' (Buck, Seifert, and Just, Consulting Engineers 1947). The results of this study and the proposal for the project were handed over to the Municipality in August 1947. The old source for drinking water for Guayaquil, the River Daule, surfaced again as the next target. As well as fulfilling an expanding role as a water source for irrigation projects in the region, the flow of the Daule was to be diverted, transformed, and commodified. Banana export earnings, combined with a reverse flow of money from the US, were welded together with the flow of Daule water to circulate through the veins of the city, transforming its geographical landscape through the upgraded circulation of water. This flow, combined with and running through physical and urban space as a material stream of H_2O, represents just one node in an articulated network of processes operating on a regional, national and, indeed, worldwide scale: flows of transformed nature, commodities (bananas) and money, transfers of capital, buying and selling of labour power (see Merrifield 1993). Once more, the city would be transformed and the political economy of the urbanization process was to be deeply bound up with the progress of the urbanization of water.

Construction of the new potabilization station 'La Toma' was initiated in 1950, and it became operational in 1951. The station captures water from the

Daule river, 26 km upstream from Guayaquil. In addition, a new main conduction pipe (43″) was constructed, which led to a new urban reservoir at Tres Cerritos in Urdesa, west of the city centre (see Figure 3.2 and 3.4). This main also serves as a distribution pipe for industries and neighbourhoods located on its trajectory. 'La Toma' initially had a maximum daily capacity of 75,000 m^3 and was designed to serve a predicted maximum population of 440,000 by 1980 (M. I. Municipio de Guayaquil 1952). Not surprisingly, within a few years the area around the new reservoir became the middle- and upper-class residential part of the city. Indeed, formal urban development, feverish land speculation, and surging land rents closely followed the flow of water. These prime urban sites, protected from land invasions, were now ripe for development and for generating significant urban land rents. The banana boom made money easier to come by and high-class urban development became a prime source of accumulation for a growing construction industry and real estate market. Indeed, one way of recycling banana rents was into the promotion and financing of urban development and housing projects through the Guayaquileño financial system and a mushrooming number of local and international banks (Villavicencio 1990: 33). As such, the emergence of a new strata of land developers, whose accumulation process was organized through the production of new urban landscapes and the seizure of rents from land development was predicated upon the extraction of banana rents and their circulation through local, national, and international processes of capital circulation.

However, the banana bonanza, combined with the modernizing politics of the national state, had begun to dismantle traditional forms of labour control and propelled unprecedented numbers of rural to urban migrants, particularly to Guayaquil. By 1957, the city's population had reached 403,000, with the result that the new water purification station was operating at its anticipated full capacity only a few years after its inauguration and the population began to suffer from major water shortages and water scarcity once again (Olaya and Villavicencio n.d.). Although the expansion and improvement of the water supply network in the central and newly developed parts of the city was in itself no small achievement, problems began to multiply. Moreover, the now rapidly spreading invasion settlements, which, ironically, were mainly built in the mangrove estuary and whose construction demanded a detailed division of labour and quite intricate engineering works to control the tidal flows of marine water, were never properly serviced. Indeed, as the banana boom drew to an early close toward the end of the 1950s, the further conquest and urban mastering of water reached a new impasse.

5.2 The closure of the frontier and the emergence of an ideology of water scarcity

In the 1950s, a new and more resistant banana variety, the Cavendish, was developed. This innovation allowed the fruit companies to switch their operations back to the more favourably located Central American locations that were closer to 'home', more reliable, and under more direct control of the US state. This new bio-engineered and phyto-technologically more demanding 'Chiquita' banana (León 1992) was heavily commercialized internationally and undermined the economic position of the traditional Ecuadorean 'Gros Michel' banana type. Only large Ecuadorean producers who were connected to international merchants and fruit companies were able to adjust ecologically and socio-economically to the requirements of the new cultivation, production, and marketing techniques. While smaller producers had difficulties coping, output continued to grow, resulting in falling international prices as world supply suddenly surged rapidly. Moreover, banana diseases had spread southwards and begun to affect Ecuadorean banana yields (Nurse 1989).

By the early 1960s, the production volume was twice the exported volume. International merchants could be more selective and demanding. Returns began to decline or, at least, their growth began to slow down. Total banana export value fell from US$88.9 million in 1960 to US$51.5 million in 1965, although it recovered by 1970 (US$94.3 million) (Tobar 1992: 238). This over-accumulation of bananas wiped out thousands of small- and medium-sized producers, who joined the ranks of the urban underclass (Báez 1985: 554). This decline broke the coastal elite's partially restored power position once again and resulted in a partial shift of capital into new areas, in particular the urban environment through land speculation and development. Ecuador's export revenues stagnated, resulting in the rapid rise of external debt and forcing the state to face up to the problem. The state responded with increasing taxation and, in order to appease the disgruntled export merchant elite, decided to devalue the currency from 15 to 18 sucre to the US dollar in 1961. The choreography of unfolding political economic and ecological changes were strangely familiar to those that had accompanied the cocoa collapse almost half a century earlier.

The collapse of the banana-based agro-export model caused a significant deterioration in the living conditions of the poor and the workers. The elites might have lost some of their luxuries, but it was the poor who bore the brunt of the devaluation processes sweeping the country. Social unrest was the inevitable outcome, particularly in Guayaquil where the consequences were more acutely felt and where social relations had a more explicit class character than in the rest of the country. On 2 and 3 June 1959, the people of Guayaquil took to the streets to protest against the onslaught on their livelihood caused by the twin processes of national austerity programmes and the international crisis of the banana industry. As in 1922, this popular revolt ended in carnage: the army was called in and more than 500 people were killed in the city (Muñoz Vicuña and Vicuña Izquierdo 1978; Ycaza 1992: 556). In 1961, another series of similar, although less violent, popular rebellions erupted. The disarray, in which both the state and the economic elites were increasingly embroiled, gave rise to the rapid growth of local populist parties and the seizure of power by the military in 1963. Needles to say, this crisis arrested the further modernization of the city at exactly the time when the urban crisis was becoming more acute.

The social unrest and the rising deficit of urban public services forced the state to seriously consider the problems of collective consumption in the marginal areas of the city. A new strategy of urban domination became increasingly apparent, with the spontaneous social mobilizations of the late fifties and early sixties gradually becoming channelled and controlled through a personalized clientelist urban political system aimed at the partial satisfaction of popular demand for key services. By now the invasion settlements were more consolidated and better organized, and represented a growing popular base for mustering political support. The emergence of neighbourhood movements and leaders enabled negotiations to take place between the local state and the neighbourhoods, which in turn strengthened populist–clientelist relations. Local neighbourhood leaders became key actors in intricate, personal, and dense networks of power and entrenched favouritism that would characterize urban politics during the subsequent decades. The developments of the urban water system, which were realized under this regime were, of course, piecemeal, uneven, ad hoc, and based on personal political favours (Villavicencio 1988; Menéndez-Carrión 1991).

The government had become extremely worried by the continuing problems of water supply in the city and its potentially dangerous effects on social stability and cohesion. Just a few days after the massacre of the protesting masses, on 12 June 1959, the national government created the 'Junta Cantonal de Agua Potable de Guayaquil' as a formally independent and autonomous organization with the remit of providing and administering 'a plentitude' of water production and distribution (Registro Official 1959). This new institution replaced the municipal 'Departamento de Agua Potable', which had hitherto been in charge of urbanizing and domesticating nature's water. The official policy was still aiming towards a water supply system providing every urban household with unlimited quantities of potable water. In fact, upholding such a view as official served (and still does) to contain a series of increasingly sharp urban contradictions. First, it appeased those that were hitherto excluded from access to water by promising a solution in the near future, without necessitating a commitment to an actual timescale. Second, it did not threaten those who already enjoyed cheap access to unlimited quantities of water as the question of equitable distribution was submerged under the rhetoric of a promised expansion of production capabilities. Third, the piecemeal improvements suggested that clientelist strategies 'worked' and reinforced the hold of both the populist parties over the urban underclass and of local leaders over 'their' neighbourhoods. Fourth, the problem of systemic water shortage was and is attributed to technical and natural constraints, with water scarcity being constructed as a matter of the unavailability of technology and water rather than as a question of the organization and management of an equitable distribution system. This politically produced ideology and practice of water scarcity defuses potential socio-political unrest over the water issue by blaming nature for the problem. A few decades later, this same rhetoric and practice of 'water scarcity' would be successfully mobilized to advocate the privatization of water supply and the introduction of market principles in the organization of the urban water sector.

In 1960, the new water company ('Junta Cantonal de Agua Potable de Guayaquil') opened a tender to undertake the first Master Plan study for the future provision of water. SEURECA, a French consultancy and engineering firm, started working on the plan in 1961. The proposals and new projects suggested by the consultants would lay the foundations for the current water supply system. Old pipes, sometimes dating back to the beginning of the century, were to be replaced while production and distribution capacity would be expanded (Olaya and Villavicencio 1990). In 1962 the 'Ley Constitutiva de la Junta de Agua Potable' was changed in order to turn the institution into an autonomous company of private right with a social objective, in other words an independent, but publicly owned, company. This allowed the water company to negotiate loans from international organizations and to contract works. The exercise was clearly also inspired by an attempt to take water provision out of the clientelist populist hands of the local authorities and political patrons. However, falling banana rents and a faltering Ecuadorean economy made loans, whether national or international, more difficult to come by, and debt servicing became more problematic.

In 1963, when the city had a population of approximately 500,000, La Toma's production capacity was increased to 93,750 m^3 per day. In 1968, a second huge main, with a diameter of 1,259 mm and a daily conduction capacity of 144,000 m^3, was built between 'La Toma' and the Tres Cerritos reservoir. These works were financed with loans and a variety of taxes, among others on hotels and imports and exports. On 4 August 1970, a Supreme Decree changed the legal position of the water authority once more, passing it once again to the Municipality. The newly created 'Empresa Municipal de Agua Potable de Guayaquil' (EMAP-G) took over from the previous Junta (Calle and Chang 1976), with a geographical remit extended to cover the whole of the Province of Guayas. In 1989, the water company was to once again switch from a municipal to a provincial body ('Empresa Provincial de Agua Potable de Guayas' (EPAP-G)). These frequent institutional reorganizations express and illustrate the web of political central/local and regional/national struggles that continue to characterize Ecuador's fragile political theatre. In 1971, the capacity of 'La Toma' was further increased by 112,500 m^3 per day, and another 60,000 m^3 per day was added in 1972. Nevertheless, these incremental alterations trailed far behind the rapid population growth of the city. By 1974, 63% of the total population of 823,000 inhabitants and 71% of the urban area had to be supplied by tankers or community taps. The now endemic exclusionary water practices laid the foundations for a thriving private water economy monopolized by water vendors whose exclusive control over the distribution of a key element of nature in the non-serviced urban areas enabled the appropriation of considerable water rents. In 1976, these 'tanqueros' sold water at 2.5 sucres (US$0.10) for a tank of 55 gallons (200 litres). At the time, it was estimated that the 69,315 families living in Suburbio paid on average 55.83 sucres per month to private distributors for their drinking water, or an annual total of 46,438,277 sucres (US$1,857,531) (Servicios Sociales del Suburbio 1976) (see Chapter 7 for a more detailed analysis).

5.3 The last gasp of the urban water dream

5.3.1 From black to white gold: petroleum urbanization and the globalization of the money/water nexus

In the 1970s, a third wave of ecological expansion and transformation once again provided Ecuador with sufficient foreign revenue to expand its economy. Indeed, the exploitation of Amazonia's huge oil reserves in the eastern part of Ecuador from 1972 onwards signalled a new wave of rent extraction and redistribution. The existing socio-spatial relations would be overhauled once more, as the actors organizing the petroleum boom produced a new set of geographical configurations. The ecological conquest of the underground of the Ecuadorean Amazonian rainforest was and is based exclusively on international petro-capital. The two earlier waves of integration in the international market place (agro-exports of cocoa and bananas) were mainly organized through the intermediation of a domestic commercial and financial oligarchy combined with international trading companies. In the case of petroleum, however, the national state would assume the role of the key interlocutor in organizing the global/local articulation of oil. The indigenous Amazonian peoples were legally dispossessed, as the state became the de facto and de jure owner of the country's 'natural' resources (Báez, cited in Farrell 1989: 146). Domestic appropriation of some of the rents took place either through the state or through spillover effects into the financial and construction industry, thereby putting the state into pole position in terms of organizing the insertion of Ecuador into the global political economic framework.

This power inevitably turned the state apparatus into a major arena for social struggle. Indeed, the oil revenues, partly monopolized by the state, resulted in continuous political power conflicts for the control, appropriation, and direction of the new investments that now became possible. In addition, the oil boom attracted considerable attention from foreign investors (mainly in services and banking), who were increasingly attracted to Quito rather than Guayaquil because of the role of the national state and its centralization in the capital city. Oil capital increased the 'bankability' of Ecuador and foreign loans started to flow back into the country.

The expansion of the ecological rent frontier was this time directed inland into the Amazon basin rather than towards the coastal regions. From 1972 onwards, oil, quite literally, flowed to the coast for export through a newly constructed 800 km-long pipeline over the Andes from Amazonia to the coastal port of Esmeraldas. In the process, crude oil is transformed into money and capital. Quito increasingly became the country's leading international financial centre, leaving Guayaquil behind in its past, but now dimmed, glory. The oil rents appropriated by the state were, in turn, reinvested with an eye toward domestic industrialization (Bocco 1987), mainly in infrastructural development ranging from port facilities and new freeways to airports and military build-up (to finance the latent conflict with Peru and appease a still powerful military command). The absence of a significant commodity-producing elite, however, turned this strategy increasingly sour (Bocco 1987).

The ebb and flow of oil turned into money would be primarily circulated back into Quito and Guayaquil. Investment flows had a clear 'urban bias', and in 1975/6, for example, 30% of state expenditures (excluding infrastructure, which alone accounted for 55% of the total budget) went to the cities (Allou 1987). Most of the infrastructure works also occurred in cities, and in Guayaquil the new international airport and the new seaport were both financed with oil revenues (Godard 1987). The spatial choreography of the combined flow of crude oil and capital implied a significant geographical transfer of value from an impoverished indigenous population in Amazonia to the urban centres. Furthermore, the appropriation of nature, socio-economically transformed into rent, also attracted international financial capital as the advanced capitalist economies switched to the Third World to recycle their accumulated petrodollars in the wake of the intensifying crisis in the core capitalist economies (Swyngedouw 1996*c*). During the 1970s Ecuador piled up a massive external debt, partially financed through the return of money originally exported in the physical form of crude oil, and promoted by foreign capital eager to recycle its accumulated financial reserves (Acosta 1990). Once again, the urbanization process was predicated upon further political ecological conquest, the third wave in less than a century.

Oil rents also served to augment the ecological basis on which the city's sustainability was predicated, such as widening the scale and scope of water control. In addition to major works of infrastructure, the construction sector generated the renovation of the 'tugurio central', the densely populated central part of the city. This in itself accelerated growth in the invasion settlements, mainly in Guasmo, as the renovated inner-city neighbourhoods pushed low-income households to the periphery of the city. The proliferating land occupations were, in turn, increasingly organized through the clientelist political municipal power structures. For example, land in Guayaquil's mangrove estuary (Suburbio, Guasmo, and Isla Trinitaria) was sold at a mere 10 sucres per m^2, a value which will be inflated away in the early eighties as the oil economy collapsed (Godard 1987; Scheers, 1991). The clientelist local authorities were instrumental in fostering the growth of invasion settlements in the geographically and ecologically difficult to urbanize mangrove estuary. The provision of services once these settlements became regularized proved to be technically difficult and extremely costly.

In August 1975, the sub-fluvial pipe bringing water from 'La Lolita' to the city was closed. The entire city now became dependent on a sole source of water, the River Daule, but oil money was invested to capture more of its water to refresh (parts of) the city. There was also increasing competition for the water from the river, as a massive irrigation scheme was initiated to encourage the colonization of the Guayas gulf area. In 1974, in the midst of the oil boom, the World Bank (BIRF (Banco Internacional de Reconstruccion y Fomento)) granted a US$24 million loan at 7.25% interest to EMAP-G, which was paid back over the period from 1975 to 1982. The interest rate is spectacularly low—and in fact negative—as dollar inflation rates rose to a two-digit level during that period. However, few of the projects covered by the loan would be implemented. The National Development Bank (FONADE, now BEDE) financed the further upgrading of 'La Toma' to a total capacity of 320,000 m^3 per day, although the conduction capacity remained the same at a maximum of 230,000 m^3 per day. Nevertheless, during very dry years, treatment capacity was seriously hampered as a result of upstream salination from the estuary and from eutrophied waters. A new, secondary capture point needed to be installed on the Pula river, which is itself a tributary of the Daule river. Work on the construction of a new and third aqueduct from La Toma to the city started in 1979, although the pipe terminated at Quinto Guayas, 10 km short of reaching the city (see Figure 3.4), at which point the new aqueduct was connected to the existing mains. The city would have to wait until 1992 before the final 10 km gap was closed and the extra production capacity reached the central urban reservoirs (Villavicencio 1992). In the meantime, EMAP-G had also constructed a 165 km-long main to connect the Peninsula with the urban reservoirs. The coastal towns of the peninsula were the favoured resort towns for Guayquileño elites wishing to escape the unbearably hot and humid conditions in the city during the rainy season, and the new aqueduct would provide these high-income temporary residents with domestic water. In the years to come, this main would turn into a prime target for water terrorists who regularly use explosives to dynamite the pipe in order to secure a wider geographical area for speculative water vending (see Chapter 8).

In 1982, The Water and Sewage division of the BIRF evaluated the effects of the 24 million dollar loan and wrote a devastating report. It concluded that the administration of EMAP-G was completely unsatisfactory; the project was hugely under-financed as a result of below-cost tariffs, the high percentage of water spillage (45%), and the enormous backlash in accounting (on average six months); and the administrative costs were extraordinarily high because of the growing number of employees. The ratio of employees per 1,000 connections is 15, while the Latin American average for companies similar to EMAP-G is around seven.

In 1978, about 760,000 inhabitants (59,766 connections) had a domestic water supply, while *c*.550,000 people (42%) were dependent on other means to access water, primarily a continuously growing number of private water vendors. About twenty companies, housing estates, and hotels had their own private wells using underground aquifer water (Gilbert-Brown and Caldwell-Ribadeneira 1980). Also in 1978, EMAP-G signed a contract with the Engineering Consultancy group Gilbert-Brown and Caldwell-Ribadeneira to work on the second 'Master Plan for Potable Water for the City and its Hinterland'. With the expectation of continued oil-based financing, the old dream of covering the whole city with a continuous supply of piped potable water was once again unearthed. This plan, permeated with the idea of the final full completion of the urban water network, was to remain the basis of Guayaquil's water politics until this day. By the time the plan was ready, however, the national and international situation was at the eve of great turmoil and change as the relatively prosperous decade of the 1970s gave way to the 'lost decade' of the 1980s.

5.3.2 Oasis and the urban desert: towards institutionalized water chaos

The 1980s was a decade of intensifying crisis formation, which would plunge Ecuador into one of its most prolonged recessions. The price of crude oil collapsed and the rents generated through the ecological conquest of the Amazon plummeted. This was followed by a deterioration of the balance of payments, while the build-up of debt demanded ever-rising service payments (Acosta 1990). The ensuing financial crisis and the intensified fiscal crisis of the state undermined the efforts to modernize the urban fabric. The collapse of the economy actually accelerated rural-to-urban migration, while the financial flows to equip the cities with the necessary infrastructure became gradually thinner. This has by now become an all too familiar story, as for the third time in the twentieth century, the urbanization process of Guayaquil spiralled out of control at exactly the time when the financial means required to service incoming migrants dried up and the elites desperately tried to protect or safeguard their own increasingly precarious condition. The earthquake of 1987, which destroyed the trans-Andean pipeline, and the devastating effects of El Niño on the process of agricultural rent extraction further accentuated this spiral of decline. This economic crisis and ecological change produced an urban crisis of gigantic dimensions: absence of investment, rapid expansion of the city, land invasions, decline in urban services, and notoriously chaotic urban management by a succession of populist local political leaders (Scheers 1991). The economic elites remained basically absent from the political terrain, while the new urban managerial elites maintained an important power position through the proliferation and cultivation of clientelist strategies.

The projects proposed in the 1980 Master Plan did not get under way until 1988, eight years behind the planned date. By that time, the urban condition had altered significantly and all new projects would increasingly run behind the realities of the urban expansion. Throughout the 1980s the water crisis would intensify, and by the end of the decade, the city was faced with the greatest water problem, if not outright crisis, in its history. In 1987, financing contracts were finally signed to begin works to improve and expand the city's water supply system for a total value of US$51 million. This sum was the combined financing provided by the Commonwealth Development Corporation (UK$4.9 million), the World Bank (US$31 million), and internal capital (US$12 million). However, in 1989, the World Bank suspended its loan because EMAP-G did not stick to the terms of the contract, which demanded managerial streamlining and improved operational efficiency. In the meantime, the national state had to intervene with a succession of emergency loans to prevent the total collapse of the system and to appease growing social unrest. In 1989, EMAP-G, a municipal organization, changed again to EPAP-G, a provincial organization, in a desperate attempt to take control over the water supply system away from the municipal clientelist political scene and to satisfy the requirements set by the World Bank. In the meantime, works had begun to expand the water pumping, treatment, and conduction capacity of 'La Toma': pumping capacity was increased to 800,000 m^3 per day and the water treatment capacity expanded from 450,000 m^3 per day to 650,000 m^3 per day by 1993. With a conservatively estimated loss of 20% (the real figure is much higher at around 40%), this would bring the total effective capacity to 512,000 m^3 per day. In 1992, the conduction capacity was increased to 640,000 m^3 per day as a result of the inauguration of the V-Guayas-Tres Cerritos aqueduct, closing the final 10 km gap from the most recent main to the reservoir. Upgrading and expansion of the new network was carried out in the central part of the city, parts of Guasmo, Suburbio, Mapasingue, and Prosperina and completed in 1994. By that time, the rate of coverage was about 68%, still leaving 32% of the population without domestic connections or community taps, with no solution in sight for the distribution problem. In 1993, the projected investment cost to achieve full distribution coverage was conservatively estimated at around US$40 million, although the geographical area that required servicing continued to grow as the invasion settlements expanded.

In the meantime, however, the obsession with producing more potable water continued unabated. In 1995, a new water treatment station, adjacent to the old one at 'La Toma', was constructed for US$75 million borrowed from the Spanish government. The total capacity of the new plant is estimated at 864,000 m^3 per day, and although it would indeed improve hours of servicing for those parts of the urban population who have domestic connections, this white elephant will once again leave the excluded at the mercy of the water speculators. As will be discussed in Chapter 7, 600,000 people were still dependent on 'tanqueros' in 1993, paying an estimated US$9.5 million for their water during that year alone. In fact, the power of the water vendors became consolidated through their continuing control over a vital commodity.

This institutional water chaos, enduring problems with water delivery, and the socially produced 'scarcity' of water began to feed a discourse of liberalization and privatization in the early 1980s, with calls for a radical shift in water governance becoming increasingly loud from the early 1990s onwards. In addition, the neo-liberal policies that swept through Latin America, combined with the structural adjustment policies and requirements for deregulation as conditions for international lending set by the world's leading financial institutions (IMF, World Bank) produced a political, social, and economic environment that pushed, slowly and originally imperceptibly, for greater private sector participation and investment in the water sector. By the mid-1990s, it became abundantly clear that Guayaquil and Ecuador would follow the clarion call of the international political and economic elites. They set off on a course of privatization, a trajectory that had now been hegemonically presented as offering the panacea for what had become a really intractable problem. This transition from a publicly owned to a privately controlled water company will be further explored in the next chapter.

5.4 Conclusion: scaling nature, scaling water

As documented in the previous two chapters, the city of Guayaquil grew on the basis of successive ecological conquests and the appropriation of rents from agricultural produce or the pumping of oil through which money was continuously recycled and nature became urbanized. The harnessing and urbanization of water inserted the circulation of water squarely into the circulation process of money and its associated power relations and class differentiations. The circulation of water and of money expressed and reflected the changing political and social power geometries and the spatial choreography of social, political, and economic power relations at a variety of articulated geographical scales: the urban, the regional, the national, and the international.

The production of new landscapes of power and the seizure of rents from land development was predicated upon the extraction first of cocoa rents, then bananas, and finally oil. Each time, social power relations became reconfigured through these combined political (and) ecological transformations. The flow of water, combined with and running through physical and social urban space, represents just one of myriad processes and interconnected networks operating on a regional, national, and indeed worldwide scale, including flows of transformed nature, commodities, and money, transfers of capital, and labour force transactions.

From the very beginning of commodified urban watering, the urbanization of water was subject to intense social struggles, as the dialectic of exclusion from and access to commodified urban services unfolded. The now endemic exclusionary water practices provided the foundation for a thriving private water economy, monopolized by water vendors whose exclusive control over a key element of nature permits the appropriation of considerable water rents. In other words, the transformation of nature is part and parcel of the power relations through which the urbanization process unfolds.

In sum, nature and society are brought together to form an urban political ecology that combines the political with the socio-economic and with the ecological in ways that render them inseparable. Urbanization, therefore, is a process infused with relations of social power, and proceeds through multiple forms of socio-spatial struggle in which (transformed and metabolized) nature takes centre stage. In the end, the water issue is part and parcel of a much wider consideration of the environmental basis of the city's existence and change. The city is a giant process of perpetual transformation of nature. Nature and society are welded together to form an urban political ecology that unites the power of socially mobilized and ecologically transformed nature with the power of money. Water circulation is, therefore, just one illustration of how nature and society become united in the production of a socio-spatial fabric that privileges some and excludes many. Water is an integral element in this process and needs to be addressed in these terms. The recognition of the social production of nature *and* of the city is essential if issues of sustainability are to be combined with just and empowering urban development. In short, the key questions running through these chapters are whose water is being urbanized, whose nature is transformed, and who has the right to the city and its waters? And these are the questions we shall turn to next.

6

The Water Mandarins: The Contradictions of Urban Water Provision

Billions of litres of water flow through the centre of Guayaquil each day, as the Rivers Daule and Babahoyo converge to form the River Guayas. Given this fact, it is perplexing to find that 35% of the inhabitants of the city do not have access to adequate and reliable water supplies and the whole city suffers from chronic absolute water shortages. In this and the next chapter, we shall explore the contradictions of urban water provision, which result in a sizeable part of the urban population, invariably the poorer end of the social spectrum, not having access to piped potable water. This situation, in turn, makes them easy victims of water speculators, the private water sellers that distribute water in non-serviced areas by means of tankers. In Guayaquil, approximately 400 tankers service a population of half a million people, or approximately 35% of the total urban population. These water-merchants buy water at a highly subsidized price (70 sucres/m^3), while they sell it for up to 6,500 sucres/m^3 (November 1993),[1] a price of up to 300 times higher than that paid by low-volume consumers who receive water from the water company. We will also explore the strategies and structure of the water company, infrastructure and investment planning, price mechanisms and control structures in the light of these exclusionary and disempowering mechanisms of the existing water system.

In short, we shall explore the contradictory dynamics of the 'Water Mandarins'. The complex networks of those that hold control over the water tap, water infrastructure, and water distribution will be excavated in order to unearth the relations of power that infuse and eventually organize the intermittent flow of water in Guayaquil. Of course, analysing the changing dynamics of water supply in Guayaquil is like trying to hit a moving target. The field research for this book was completed in 1998. Since then, the public water company has awarded a concession to International Water Services, a Dutch-based subsidiary of Bechtel and Edison Spa, to operate, administer, and expand Guayaquil's water and sewage services and infrastructure (see below). The private operator started officially in April 2002, and the implications and effects of

[1] The exchange rate in November 1993 was *c.*1950 sucres for 1 US dollar.

this part-privatization are not yet clear or known. Nevertheless, it is safe to say that Guayaquil continues to suffer from major water problems and will continue to do so in the foreseeable future. In August 2002, the private water-vending trucks and their sales practices remain an integral part of daily life in the poor settlements of the city.

6.1 The rise and fall of the urban public water company: the route to privatization

In the late nineteenth and early twentieth century, the circulation of urban water in Latin America tended to follow a route similar to that taken just a few decades earlier in Europe (see Goubert 1989). By the mid to late nineteenth century, urban water engineers aimed to provide full mastery and control of water for urban purposes and these mega water work systems became the hallmark of urban water provision throughout Latin America until this very day. Western capital, engineers, and technology constructed most, if not all, of Latin America's urban water and sewerage systems. However, the urbanization of capital in the context of peripheral modernization quickly showed tensions and problems with respect to the urbanization of water.

In most Latin American countries, urban water production, conduction and distribution was until very recently (and still is in many places) a monopolized and highly subsidized public sector activity, institutionalized in national, regional, or municipal water companies (Saenz 1988; Faudry 1987; Uclés 1991; United Nations Centre for Human Settlements 1991). The origins of public urban water utilities in the second half of the nineteenth century were part of the bourgeois-liberal revolutions that swept through Latin America. In Ecuador, the Liberal Revolution of 1895 headed by Eloy Alfaro provided a strong impetus for major infrastructure works, of which urban water supply became an important component. As detailed in the previous chapters, the growth of the city and the changing urban class composition resulted in a series of contradictions which were contained by means of public sector organization of urban services and infrastructure. The state provided urban services for the urban elite, while subaltern classes were systematically excluded from both political power and urban services.

Initially, the urbanization and domestication of water kept pace with the accelerating urbanization of capital. From the very beginning, the control over and harnessing of water was inscribed in the political economic struggles that underpinned Latin America's urbanization processes. However, the urbanization of water was also subject to intense social struggles from the very beginning, as the dialectic of exclusion from and access to commodified urban services unfolded. Throughout the twentieth century, the state tried in vain to keep urban infrastructure expanding in pace with accelerating urbanization.

However, the perpetual budgetary problems of the state meant that state investment was continuously in peril, resulting in domestic underinvestment combined with a chronic dependence on external financing. The contradictions of peripheral urbanization soon intensified the conflict over the control for water and resulted in a systematic exclusion of the new poorer urban residents from easy and cheap access to the (limited volume of) available potable water. In the next section, the characteristics and implications of these tensions as they are expressed in the structure of urban water provision will be further detailed. In Guayaquil, the acceleration of the urbanization process from the early fifties onwards rapidly outpaced the expansion of the water production and supply network, with the gulf between need and available service widening over the years.

As the urban condition deteriorated and the crisis of the state deepened throughout the 1980s, the public provision of collective goods came under intense pressure. From below, residents increasingly protested the state's inadequacy in providing basic services; from above, fiscal constraints, liberalizing tendencies, and the tensions within the Keynesian model of state investment forced a fundamental reworking of the relationship between state, city, and the provision of 'collective' goods. From the late 1980s onwards, a series of far-reaching reforms in the water sector paved the way for the part-privatization of Guayaquil's water services. The municipal company that had been formed in 1976 (EMAP-G) was turned into a provincial company in 1989 (EPAP-G), a change, which would later be seen as the first step towards the eventual privatization. The latter was finally achieved in 2000. We shall now turn to analysing the multiple contradictions of the system of public or collective water provision as they became part of the institutional and organizational structure of Guayaquil's water supply system.

6.2 The contradictions of urban public water provision

6.2.1 Chronic problems of financial deficits

The planning, programming, administration, and implementation of urban waterworks was caught up in a series of problems associated with the nature of the peripheral urbanization process. The state-controlled urban water bodies attempted to contain the contradictions between the demands of the urban bourgeoisie and those of recently arrived immigrants and poor urban residents. This can be exemplified by (a) the water tariff and price structure; (b) the problem of unaccounted water use; (c) the internal administrative functioning of the water institutions; and (d) the organization of waterworks planning and implementation.

Latin American urban water institutions are often faced with the problem of an operating situation that generates structural deficits. This is quite different from a subsidized model in that such a system would still run on a standard balanced account management. A system running structural deficits operates on ad hoc, piecemeal, and 'emergency' interventions, loans, and subsidies from the national state or international lending bodies. From the very beginning, the high cost of urban engineering works required massive (usually external) financing, while the political economic forces in the city demanded low water prices. This resulted in chronic losses for the water utilities.

In 1988, more than half of the urban water companies and all of the big city companies in Latin America were running sizeable deficits (Faudry, Ortiz, and Fandino 1991). A review of projects financed by the World Bank showed that the effective price charged for water is only about 35% of the average cost of supplying it, while the utilities generated only 21% of the projects' total costs (World Bank 1992). The 'Empresa Provincial de Agua Potable' (EPAP-G) of Guayaquil, for example, recorded an annual income from water related activities of 14.2 billion sucres in 1991, compared to a total operating expenditure (excluding amortizations) of 18.7 billion sucres. This suggests that the operational loss alone of EPAP is 24.1%. In 1990, the effective running cost (excluding capital stock depreciation) of producing a cubic metre of distributed and accounted for water was 170 sucres, although the price to low volume consumers (<15 m^3/month) was 30 sucres and to the 'tanqueros' or water vendors it was 70 sucres. This chronic deficit leads to perpetual underinvestment, problems with the maintenance of existing infrastructure, and a lack of funds to expand the network into newly urbanizing areas.

Water tariffs are often set on the basis of fixed lump-sum payments. Accurate accounting of water use is hampered by the absence of water meters, deficient maintenance, and/or insufficient meter readers (Faudry, Ortiz, and Fandino 1991). In addition, tariffs are usually set well below production and overhead costs and do not take investment costs into consideration. Water companies operate within a context that puts social, economic, and political premiums on keeping water prices as low as possible, with the aims of combating inflation, maintaining social peace, and satisfying the demands of large industrial and commercial consumers for easy, cheap, and highly subsidized access to great (if not unlimited) quantities of potable water, even though a substantial share is used for irrigation purposes, filling swimming pools, and the like. This perpetuates the chronic deficit of the public utilities (Caravedo 1986; Coing and Montano 1986) and further jeopardizes both the maintenance of the existing structure and the implementation of new investments to keep pace with urban expansion. However, the redistributive aspects attributed to the provision of subsidized water are highly questionable under conditions in which nearly all poor urban residents are excluded from the official supply system. In fact, there is a *negative* redistribution (Estache, Gomez-Lobo, and Leipziger 2001). Non-connected (poor) urban dwellers pay extraordinarily high prices to the 'water speculators' for their water because the public system cannot deliver its services in a comprehensive way, at least partially because of chronic deficits. Nevertheless, these same people contribute to the financing of the water utility's losses and the debt servicing. Within the limited context of the official system itself, some redistribution does occur which seems to benefit smaller consumers (Corchuelo Rozo 1981). On the whole, however, the chronic deficit and underpricing of water results in a massive transfer of income to commerce, industry, and the more wealthy private consumers (Faudry, Ortiz, and Fandino 1991; Caravedo 1986).

In addition to the subsidized price, and partially as its consequence, a sizeable volume of water is not accounted for. In Lima, this represents 62% of the supply (Caravedo 1986), in Guayaquil the figure is 65%. In Bogotá the rate of increase of the volume of potable water produced is higher than the rate of population growth, although the amount of water actually charged for has increased at a much slower rate (Jaramillo 1988). Four main factors are responsible for these problems with accounting for water:

(1) physical losses as a result of lacking or insufficient maintenance;
(2) free deliveries;
(3) underestimation of water use due to absent, unchecked, or ill-functioning meters;
(4) illegal and fraudulent connections by industry, households, and private water sellers.

Moreover, Zolezzi and Calderon (1985) report in their study of Lima that, contrary to popular belief, the level of unpaid water use is considerably higher in the central upper-class parts of the city than in the lower-class periphery. Evidence from Guayaquil indicates a similar situation. Data presented in Table 6.1 show that the more affluent northern and central parts of the city receive respectively 70% and 26% of the available water. Both physical and commercial losses are very high in these parts, compared with the poorer southern part of the city. The latter receives only 4% of the available water (for an average of four hours per day), but pays for 38% more water than it actually receives. As a result of the absence of meters, the residents in this part of the city are charged on the basis of *estimated* water use, estimates which are systematically higher than the actual water use. In sum, in the poorer parts of town there is 38% over-accounting, while the richer areas do not pay for 19–20% of the water they actually receive. The combination of a subsidized price with systematic under-accounting leads to a double advantage for the upper-class areas in comparison with the poorer ones. The data also challenge the deeply held belief that it is the poor in marginal settlements who are mainly responsible for illegal water tapping and water theft.

In short, the combination of below-cost pricing and the substantial amount of water unaccounted for serves particular interests. These are further accentuated in a context of widespread and deeply rooted clientelist political traditions (see, for example, Bataillon and Panabière 1988; Menéndez Carrión

Table 6.1. Water consumption, distribution, losses, and accounted-for water in Guayaquil, 1990

Sector	North	Central	South
Number of inhabitants	421,214	422,985	272,393
% of water supplied	70	26	4
Water/capita/day (litres)	307	160	43
% of accounted water	29	48	1.1
% of physical losses	53	32	27.4
% of commercial losses	19	20	–38

Source: EPAP (1991*a*).

1986). Any attempt at reducing financial losses by increasing prices or improving accounting practices would be deeply unpopular among the upper classes. In addition, implementing such price policies might lead to social unrest in the popular settlements, while improved policing of illegal connections or a more repressive stance on water theft might deprive those for whom this strategy is often the only possible means of securing affordable access to at least some quantity of potable water. Of course, these are exactly the strategies that privatized water utilities are likely to pursue after privatization.

The clientelist-populist nature of many Latin American urban polities, a condition with a long tradition in Guayaquil results in the water companies having a very low level of productivity, a situation which is exacerbated when combined with low labour costs. The ratio between the number of employees per 1,000 connections is generally high (on average about eight) (Montano and Coing 1985), leading to Kafkaesque bureaucratic procedures and working environments. The administrative overkill weighs heavily on the budget, and the budget is further stretched by insufficient maintenance, which increases long-term costs. In addition, the chronic deficit necessitates external financial inputs, resulting in a situation of perpetual debt, the servicing of which further increases the already sky-high budget deficit (Jaramillo 1988).

Moreover, inadequate or absent urban planning pushes overhead costs further upwards. Water engineering systems follow urbanization patterns rather than guiding them. As many of the spontaneous settlements take place on marginal land that is difficult to access and/or poses serious technical or engineering difficulties, the cost of bringing water to these communities is often very high (CIUDAD 1988). The absence of actual or prospective land rents of any significance in these 'marginal' areas turned them into prime 'invasion' sites. However, extraordinarily high water rents rapidly negate the benefits of these low land rents as the new urban residents become subject to the monopolistic practices of private water vendors (see Chapter 7).

In sum, the internal administrative and operational organization of many Latin American urban public water utilities reflect strategies to contain the contradictions outlined above, which then begin to work themselves out within the structures of the water institutions. This, in turn, aggravates an already acute urban socio-ecological crisis.

6.2.2 Structural dependence on outside financing

The precarious budgetary conditions of the water companies make them dependent on external sources of financing to make up for the deficit and/or to invest in new ventures. National subsidies or loans are rarely sufficient, so international financing is usually required to maintain or expand the system. Montano and Coing (1985: 22) hold that the model of Third World urban water supply is based on external (international) financing, and argue that this condition accentuates centralizing tendencies and favours large-scale engineering projects. Consequently, the management and technology of urban water provision is deeply caught up in the political economies of national states, their entrance to and position within the global division of labour, and the strategies of the chief international private and public lending agencies.

The dependence on national finance takes the form of annual subventions and/or ad hoc and often *post hoc* 'emergency' subsidies or loans. This dependence leads to a situation in which urban water supply is caught in the political debate on national strategies on the one hand and the national economic conjuncture on the other. In addition, it ultimately fuels inflationary pressures (Bataillon and Panabière 1988). The money available for collective consumption investments has shrunk considerably, particularly during the period of intense financial crisis in Latin America over the past 20 years, and not least as a result of the massive debt burden. Annual inflation leapt well into two-digit (and sometimes three-digit) figures, leading to extreme attempts at achieving fiscal stability, such as the full conversion of Ecuador's economy to the US dollar in 2000 (*dolarización*). Not only did national expenditures fall in real terms during the 1980s and 1990s, but also the economic crisis, combined with the imposed austerity recipes from international lending agencies and a decisive shift towards deregulation and more market-oriented national economic policies, diverted capital away from social or collective investments (Acosta and Moldonado 1992). The object of this structural adjustment policy was to switch capital from the public to the private sector in a (largely vain) attempt to combat inflation and to put the national economy back on the rails of a re-invigorated export dependency. Below, we shall discuss how the conditions set by international agencies paved the way for the privatization of Guayaquil's water services. In spite of the 'International Drinking Water Supply and Sanitation Decade' of the 1980s and more recent resounding international declarations (see Vásconez 1992), water infrastructure investment has slowed down

Table 6.2. Evolution of the debt position of the Empresa Provincial de Agua Potable, Guayaquil, 1980–1992

Year	Total debt in US$	Debt service	
		in US$ (BIRF)	in sucres (BEDE)
1980	42,735,010	1,791,466	67,437,151
1981	40,943,544	1,935,519	74,810,681
1982	39,008,025	1,922,743	65,053,380
1983	37,085,282	1,650,000	87,682,850
1984	35,435,282	1,969,249	111,751,389
1985	33,466,033	1,886,872	165,291,126
1986	31,579,161	2,197,767	168,166,139
1987	29,381,394	2,247,206	193,840,502
1988	27,134,188	3,207,354	268,844,076
1989	23,926,834	2,457,825	398,173,198
1990	21,469,009	3,003,236	436,383,221
1991	18,465,773	3,000,994	444,429,818
1992	15,464,779	2,969,531	415,749,075

BEDE: Ecuador's National Development Bank (Bando Ecuadoriano de Desarollo Economico).

Source: EPAP-G.

and the supply of potable urban water has decreased in many Latin American countries. In 1991, the Ecuadorean national government had to provide EPAP with 'Aportes Emergentes' (emergency loans), totalling 35.5 billion sucres, in addition to 3 billion sucres from transfers received from Telephone Company (IETEL) contributions and another 1.3 billion sucres in subsidies from a series of other sources and special taxes. Table 6.2 shows the expansion of EPAP's debt between 1980 and 1992, a period of intense economic crisis. Most of the loans for repairing, improving, or expanding the existing production and distribution capacity were external, coming from either the World Bank (BIRF) or from bilateral sources of financing.

The high levels of debt, exacerbated by rapidly rising interest rates, made self-financing of new investments virtually impossible. In addition, the growing international difficulties associated with Ecuador's problematic debt repayment history made international loans less easy to come by (Acosta 1990). This, in turn, slowed down the further expansion of the water system and, consequently, perpetuates the structure of exclusion from access to water. It also allows local elites to blame international lending agencies for the continuation of the existing exclusionary water distribution system.

The inadequacy of local finance necessitated perpetual appeals to international financing, which grew from funding 30.2% of water supply infrastructure development in the 1970s to 43.5% in the 1980s. During the 1980s, the International Development Bank provided about half of this financing, and the majority of the rest came from the World Bank, while bilateral financing (including by the European Union) represented no more than 5% of the total (Faudry, Ortiz, and Fandino 1991). However, interest rates and Ecuador's debt burden shot up over the same period, making international capital (which is always a rather rare commodity to come by for state-controlled collective consumption investments) even more difficult to obtain. Moreover, the continuous dependence on foreign capital aggravated the national debt problem and perpetuated the infernal vicious spiral of the country's external debt burden (Cuervo *et al.* 1991).

In recent years, international lending bodies have turned their attention somewhat from a focus on large new infrastructure works to institutional streamlining, administrative and operational efficiency, and maintenance (Zavala 1990). This, of course, is also considered to be a necessity to pave the road for privatization. However, loans also became subject to more stringent managerial and productivity conditions, which many water companies find extremely difficult to adhere to in the short run as a result of the social and political forces under which they operate. In addition, they are acutely aware of how such conditions facilitate privatization, a prospect many workers (whether high ranking or rank and file) do not look forward to.

For example, in 1988, the World Bank put a loan approved in the early 1980s to Guayaquil's water utility on ice because EPAP-G could not adhere to the Bank's strict managerial and organizational conditions. These conditions included reducing the amount of water lost through damage and theft, reducing the workforce of the utility and increasing the price of water. For reasons associated with the nature of the local state (see above) and the widespread clientelist traditions of Ecuador's local politics, a rapid implementation of these conditions is impossible, as they demand radical changes in a very well established and deeply entrenched political practice. The World Bank released the loan only in 1994, after a major institutional overhaul of EPAP's internal organization (see also below).

The dependence on international finance also has a bearing on the type of technological water control pursued (Brustein 1988*a*). The requirements of international lending agencies make project execution subject to international tender. Bilateral loans are usually subject to the use of engineering services and procurement of equipment from the lending country. These mechanisms privilege new infrastructure investment over institutional or organizational improvements and maintenance projects. They also contribute to a costly heterogeneity of systems and technologies, and close off the market from domestic suppliers. Such imposed technical choices pose problems in terms of both the qualifications of local engineers and in terms of maintenance. Costly imports of foreign components and dependence on external expertise are the most important problems associated with these practices. These processes further feed the 'productivist logic' that characterizes most water supply utilities, a problem that we will turn to in the next section.

6.2.3 The 'productivist logic'

The deadlock in tariff increases, the negative returns on water sales, the historical preoccupation with massive engineering structures for the production and transmission of water, the ideological bias towards providing unlimited quantities of water to industry and higher classes, the chronic shortages of water as a result of accelerating urbanization, the dependence on external finance, the technological bias inherent in international lending—all these combine to result in a preoccupation with the production and transmission of potable water and a negligence of maintenance, accounting, distribution and consumption, not to speak of sewerage and the treatment of waste water. This, in turn, perpetuates the systematic exclusion of large parts of the population from access to the available water. These processes also enable private water distribution monopolies to prosper and to consolidate their power by means of water speculation and monopoly rent extraction.

In Guayaquil, the average production and supply capacity of the existing facilities would allow for a daily per capita consumption of 220 litres. Current daily consumption varies between an average of 307 litres per capita in the well-to-do parts of the city to less than 25 litres per capita for those who are supplied by the private water sellers. Compared with an internationally accepted standard of 150 litres per capita per day, Guayaquil ought to be in a position to provide every citizen with a sufficient supply of potable water. The key issue, therefore, is one of distribution of available capacity rather than an issue of absolute scarcity. The scarcity that is actually felt by the residents of some parts of the city is socio-politically constructed rather than produced by environmental or technical constraints.

This inequality in terms of access and consumption serves as a powerful instrument to legitimize a productivist ideology and helps to avoid discussing, let alone, tackling the thorny distributional issue. Indeed, the shortages in production and conduction capacity to meet the potential demand for water (which is usually estimated on the basis of a combination of average and projected uses of current water customers with internationally accepted recommendations, which are largely extrapolations of current water use in Western societies) feeds the productivist logic of water utilities and their political and economic representatives (Marvin and Laurie 1999). The proclaimed panacea for an apparently 'technically' and 'rationally' measured insufficient supply (itself a highly contested concept) is the expansion of production and conduction capacity. The latter increasingly runs behind the realities of contemporary urban life. The growing mismatch between available service and popular need, in turn, further reinforces the call for more production, more investment and new facilities of the traditional kind. As such, the exclusion from water or the highly unequal access to water provides a strong argument to those commanding water production and supply to perpetuate and strengthen a system that is fraught with actually producing the very exclusionary practices it set itself to solve in the first place. In fact, even the 'water speculators' applaud these policies as they provide the best assurance to safeguard their monopoly rent extraction practices. Obviously, as long as the state and its elite administrators consider private water vending to be only a temporary 'emergency' condition that will disappear as soon as the 'master plans' are implemented, there is no need to regulate, control, or institutionalize this alternative water delivery system in the interest of the poor. This objective alliance between the 'formal' and 'informal' water sectors is further developed and strengthened through allegedly reported direct relationships between the water administrators and the 'water speculators'.

Until recently, the transformation of nature's water was a wholly public activity, although the urbanization of water is organized through a mixture of public and private systems. The public system, however, supplies urban elites and powerful enterprises, while the poor are served predominantly by the private system. The elites enjoy subsidized water supplies, but the poor are caught in the monopolistic squeeze of the private vendors. For example, EPAP's total revenues from water sales in 1991 amounted to US$12.35 million, while the estimated total income of the private water vendors (who distribute not more than about 3% of the available water—12,000 m^3 per day) exceeded US$7.2 million (Swyngedouw and Bovarnick 1995). Moreover, the productivist bias allows water companies to concentrate on technical engineering issues (which are, of course, not neutral in their social content and effects) and to avoid the thorny and politically controversial issue of just water distribution. Demand is considered to be a given. It is not an issue for which choices have to be considered, and difficult questions such as equitable management and the inequality of access do not have to be addressed.

The above set of mechanisms leads to an overwhelmingly technocratic perspective and gives supply priority over demand in the chain of circulating water. The 'productivist logic' puts a premium on costly, new, centralized, and large-scale production systems at the expense of distribution and maintenance. The use of conventional technological systems to cover the urban area in a comprehensive manner would be too expensive, a situation that consequently induces segregation in terms of access to water. The long-term financing of such giant conventional engineering works precludes the possibility of wider coverage of water supply in the short term, which, combined with lagging provision of sewerage, further consolidates already pervasive patterns of residential segregration (Goldenberg 1981; Knaebel and Leme 1985). In 1992, Niemczynowicz estimated that '[a]pproximate cost estimates for the complete provision of water supply and sanitation in urban areas of the developing world by the year 2000, calculated according to the UNICEF model, amount to US$357 billion, taking into account that only 50 per cent of this amount would be used for high-cost technology' (see also Christmas and de Roy 1991). According to the 'New Delhi Statement' (1991), meeting all demands within the water and sanitation sector by 2000 would have required an increase of investments from US$10 billion per year in 1991 to US$50 billion per year by the year 2000, using conventional approaches. We know now that this target was not achieved. At the Environment Summit in Johannesburg in 2002, the intention to halve the number of people without access to safe water by 2015 was repeated. However, without political will and extraordinary financial inputs, this will once again remain a distant dream.

The narrow productivist and commodifying logic of urban water supply also excludes considerations of water resource management and focuses attention solely on the production of potable urban water without considering alternative or multiple uses, recycling and treatment, or the sustainable management of 'natural' water resources. In addition to the obvious problems this causes in terms of ecological management and the sustainable use of water, this also intensifies conflicts with other uses of water such as agricultural irrigation and hydroelectrical production.

6.2.4 The rush towards privatization

Over the past decade, privatization of public services has rushed ahead, particularly in Latin America (see Table 6.3). By 2007, it is expected that 60% of Latin America's water and sewerage services will be operated privately. This spectacular growth (see Table 6.4) coincided with a worldwide rush towards deregulation, reduced state involvement in the economy, retrenchment of public service delivery, cutbacks in public spending, fiscal restraint, and reduced external lending. The contradictions of public water provision produced a series of tensions, problems and, eventually, difficult to manage conditions that paved the way and cleared the obstacles for private investment in the water

Table 6.3. Proportion of water and sanitation services privatized, 1997 and 2010 projected

Region	% Privatized, 1997	% Privatized, 2010	Value of privatized market (US$ billions)
Western Europe	20	35	10
Central and East Europe	4	20	4
North America	5	15	9
Latin America	4	60	9
Africa	3	33	3
Asia	1	20	10

Source: www.thewaterpage.com (5 Sept. 2002).

Table 6.4. International corporate private investment for water and sanitation in developing and transition countries, 1984–1997

Year	Number of contracts	Increase (%)	Value (€ million)	Increase (%
All Developing Countries				
1984–90	8		300	
1990–7	97	1,137	25,000	7,900
Breakdown by region, 1990–7				
East Asia	30		12,000	
Eastern Europe/Central Asia	15		1,500	
Latin America/Caribbean	40		8,300	
Middle East/North Africa	4		3,300	
Sub-Saharan Africa	8		37	

Source: Elaborated from Department for International Development (2000)

sector (Crespo 2002*a*). The reasons for and motivations behind privatization, as summarized by Castro (2002) (see Tables 6.5 and 6.6), do indeed hold for the case of Guayaquil, despite the arguments that militate against privatization (see Table 6.7) (see also Gleick *et al.* 2002). Of course, there is nothing inevitable about privatization, which is, in fact, a carefully orchestrated process by the state, usually in conjunction with international organizations, that shapes the political, economic, and institutional conditions that eventually lead to a wholesale overhaul of the public services. In addition, selected media join the chorus to chant the virtues of privatization and hail this as the panacea for all ills. Moreno Mendoza (1998) argues, for example, in *Vistazo*, a local magazine, how privatizing water services will simultaneously deal with a recalcitrant labour force in the water company and solve the problem of informal 'tanquero' water vending in the suburban areas. The recent institutional history of Guayaquil's waterworks illustrates the creeping, but targeted, process that would ultimately result in the privatization of the city water and sewerage services.

In 1987, the World Bank had approved a US$31 million loan for the improvement of water services. This was subject to a series of conditions, such as institutional streamlining, the integration of water and sewerage services, and improved operational efficiency, and was suspended in 1989 after EMAP failed to fulfil these. From that moment onwards, the process of institutional change accelerated. The change of EMAP-G, a municipal organization, to EPAP-G, a provincial organization in 1989 was the first step in adhering to the conditions set by the Bank, accompained by a change in leadership in order to facilitate cleaning up the internal administrative chaos. In addition, it paved the way for the integration of water and sewerage services as the public sewerage company had been provincially organized for a long time. In 1994, the

Table 6.5. Factors of public sector inefficiency

GENERAL TO ALL TYPES OF PUBLIC COMPANIES	SPECIFIC TO WATER SECTOR COMPANIES
Economic/financial Private companies that had to be rescued by the public sector owing to financial troubles Shortfall of capital Inadequate funding provision to meet agreed targets Propensity to incur in excess indebtedness Obstacles to access capital markets Lack or irregular valuation of fixed and operational assets Poor planning in the allocation of resources and priority setting Poor management of targeted subsidies	Economic/financial Centralized government that prioritizes expansion investment forgetting operation and maintenance Socio-economic characteristics of the customer base that limit revenues or induce social and labour problems Financial constraints to meet demand growth (for water supply, electricity, etc.) Spiralling operational and maintenance costs Inadequate tariff structures
	Technical/environmental High variations in sources and uses of water High water losses Droughts or environmental change that affects the availability of water resources
Policy/organizational Inadequate legal framework Over dimensioned structures Prevalence of short-term approaches to long-term requirements Lack of precision and consistency in objective setting Lack of managerial experience Excessive turn around of personnel owing to political interests Lack of monitoring and supervision of performance (internal and external) Excessive bureaucratization Corruption Deficiency in the personnel's aptitudes and attitudes (lack of commitment; low motivation (e.g. owing to low salaries)) Limited professionalism Inadequate career structures	Policy/organizational National framework prohibits or constrains business orientation in running the utilities Institutional complexity and overlapping of responsibilities (national, regional, and local levels)
Political Government interference in the management process Influence of external (mainly political) factors on decision-making Politicization of managers	Political Legal, political, and social limits to tariff increases

Source: Castro (2002).

water and sewerage services were integrated into one company, the Empresa Cantonal de Agua Potable y Alcantarillo de Guayaquil (ECAPAG), in fulfilment of another condition set by the international lending institutions. A US$40 million loan from the Inter-American Development Bank to the

Table 6.6. Factors promoting privatization

Economic/financial
The economic and financial needs of the country in the face of scarce resources and increased demand
Reducing the financial burden of the public sector
Capacity to attract foreign loans
Increased productivity gains expected from introducing competition
Promotion of economic growth expected from private investment
Assistance to development of capital markets
Generating resources that can be applied in cross-subsidization to fund other projects (e.g. network expansion, waste water treatment, etc.)
Reduction of taxes through raising revenue from water
Technical/environmental
For expanding the service to unserviced areas
For quality improvements (e.g. new water treatment technologies, etc.)
Tackling the increased complexity of water supply activities and their environmental impacts
Policy/organizational
Reducing the administrative burden and direct responsibility of the public sector
Improvement of the capacity for response to local needs
Transparency of information (on funding sources, guarantees, risks, conditions)
Increased private sector confidence inducing investment and repatriation of capital
Political
Pressure and conditioning from multilateral financial institutions
Pressure from powerful interest groups that may benefit from privatization
Expected enhancement of the government's credibility (internally and externally)

Source: Castro (2002); see also Gleick *et al.* (2002).

government of Ecuador specifically required the privatization of ECAPAG in order to 'improve the water and sanitation services to the city of Guayaquil'. In fact, the loan included provisions to grant a long-term concession to the private sector to 'promote greater efficiencies of these [water and sewerage] services and investment in the system' (Inter-American Development Bank 1997). The loan was granted to enable ECAPAG to undertake technical, legal, and financial studies to prepare bid specifications to award the concession. ECAPAG would be reorganized to function as an oversight and regulatory agency (with a massively reduced workforce), while the private sector would be responsible for operating the system. Only half of the loan (US$19.8 million) was earmarked for financing high-priority rehabilitation work to prevent further deterioration of existing infrastructure. The loan has a 25-year term at a variable rate, in 1997, of 6.9%, and local counterpart funds totalled US$10 million. International bids were solicited, but despite the fact that Suez-Lyonnaise and Thames Water pre-qualified, International Water was the only company to bid. In January 2001, ECAPAG confirmed that International Water had been awarded a 30-year concession to operate and administer the city's waterworks.

Table 6.7. Factors discouraging privatization

Economic/financial
Lack of interested and reliable investors
Unreliable economic environment (e.g. inflation, volatile exchange rates threatening returns on investment, etc.)
Poorly developed capital markets
Prospect of price increases or excessive profits
Potential creation of monopolies
Difficulties in collecting water payments, uncertainties on levels of return to be achieved
Higher than average rates of return requested by the private investors
Fiscal deficits that limit financial capacity
Technical/environmental
Unreliability or inadequacy of resources availability, geographical obstacles, etc.
Policy/organizational
The complexity of the preparatory work (legal, institutional, economic, political, etc.)
Legal or constitutional prohibition
Inadequate or unpredictable institutional and policy environment
Inefficient or nonexistent regulatory structures
Political interference
Low motivation of the staff, salary problems
Corruption
Socio-political
Weight of tradition (government as provider)
Government interest in keeping control over services (for social, political, or economic reasons)
Uncertainty about privatization prospects in the face of inconclusive evidence
The scope of the changes involved
The protection of social equity aspects (distribution issues, welfare, health, etc.)
Opposition interest groups that have stakes in public enterprises
Political opposition
Private sector distrust of long-term viability (e.g. fears of re-nationalization)

Source: Castro (2002).

A local subsidiary, Guayaquil Interagua C. Ltda., was established through an investment company (International Water Services B.V.) based in the Netherlands.

International Water (IWL) is a truly global company. It originated in the early 1990s as part of a series of partnerships between North West Water (NWW) (now United Utilities (UU)) and Bechtel (50%), the US construction company. Bechtel purchased NWW's engineering division and created a joint water venture in the USA, US Water. International Water was created as a joint venture operating internationally outside the USA or the UK. UU sold their share to Edison Spa, an Italian company, in 1999, but remained as the agreed 'operating partner' for International Water. IWL is globally active, with operations in, among others, Melbourne, Manila, Mexico City, and Sofia. It was also the contractor in the 'failed' privatization of Cochabamba's water supply system (*Financial Times* Global Water Report 2000*a*; 2000*b*; Crespo 2002*a*). All former ECAPAG workers were dismissed and selectively rehired by IWL. After long negotiations Guayaquil Interagua agreed to contract most of the former employees, to provide training to those hired, and assistance and training for those who would not be re-hired. Both training programmes were funded by an Inter-American Development Bank loan, through its Worker's Transition Program. However, the company broke the agreement and contracted only about 20% of former employees. Workers are now taking legal action against the company (Hall and Lobina 2002; Acosta 2002). The World Bank's Multilateral Investment Guarantee Agency (MIGA) signed a US$18 million guarantee in March 2001, offering protection for the investment against the risk of expropriation, war, and civil disturbance, and also covers a performance bond. This was the first time that MIGA had guaranteed a water project (Multilateral Investment Guarantee Agency 2002), and it provides financial security from risk for the privatized water concession of IWL. Of course, ultimately, this risk is also covered by the public purse. IWL had a disastrous experience in Cochabamba, Bolivia, where the privatized company was re-socialized after massive public protests, which took on a national significance and almost toppled the government, with a number of people being killed by the police (Gleick *et al.* 2002). After this, it became increasingly necessary for the World Bank and other organizations to insure private investments in the water sector against such 'eventualities' that might jeopardize the long-term profitability of the investment.

In the privatization contract, IWL offered to create 55,238 new connections in the first five years, achieve 95% service coverage by year ten of the concession, and invest US$520 million in the last 25 years of the period (*Financial Times Information*). It would be more than a miracle if they were able to achieve this. Operations of the private company started in April 2002. On 4 September 2002, *El Universo* reported that only 30% of the customers pay their water bill. The price charged by the 'tanqueros' for a tank of 200 litres of water stands now at US$0.80. After more than 100 years of some sort of public water supply, the Guayaquileño's are now drinking water supplied for profit. The water of the River Daule is now flowing through privately organized water systems that turn water into profits for globally organized private companies.

6.3 Conclusion: whose water?

In sum, the position of the water supply system in much of Latin America is caught between the political economic forces operating at the urban level, the struggle at the level of the state with respect to the allocation and distribution of resources, and the dependent position of these countries within the international division of labour. The above analysis does nevertheless suggest that the operation of the hydraulic system cannot be analysed independent from its organizational and institutional configurations and the relations of power that structure them.

Up to 50% of Latin America's urban population has no, limited, or only extremely difficult and costly access to water, nature's most important element. We have argued that the characteristics of public water provision prevailing in most Latin American countries produce, at least in part, these exclusionary conditions of water accessibility. These practices are the result of a series of interconnected processes:

1. A particular form of controlling nature by means of large-scale, hierarchically organized and controlled, and technology intensive infrastructures. These forms of engineering nature imply a set of social power mechanisms aimed at providing unlimited quantities of low-cost water to the urban elites.
2. A management system suffering from chronic deficits, as a result of less-than-cost-pricing. This prevents auto-financing of maintenance and service expansions, and results in a position of dependence on international lending bodies.
3. The dominance of a productivist logic which results in a disproportionally important emphasis on water production and conduction and a relative negligence of maintenance, organizational reform, sewage treatment and, in particular, the thorny and controversial issue of just distribution.
4. The resulting tensions and contradictions cleared the way for the privatization wave of water and sewerage systems.

In short, the unequal access to water and the exclusionary practices of local water politics perpetuate and strengthen a system, which, in the end, is the result of a political economic organization whose official aim is to eradicate the exclusionary practices it produced in the first place. The issues raised in this chapter point to key questions with respect to sustainable urban development in the Third World. Water is essential material for maintaining bodily and social life. Yet, the political ecology of urban water provision does exclude large segments of the population from nature's water. Therefore, the issue of sustainable urban development must raise the questions of 'Whose water?' and 'Whose city?'. The management of nature's water and the management of the 'urban' as a process of nature's transformation must ask questions about a just distribution of the available resources. And this requires a greater democratization of exactly the socio-natural metabolic processes through which nature becomes urbanized.

Elements for a sustainable, emancipatory, and empowering urban water circulation system include political, ecological, managerial, production, distribution, and supply considerations. In the first instance, it is imperative that the productivist logic and the ideology of the proclaimed need to provide unlimited quantities of water to every urban resident is replaced by a water management system based on a more equitable distribution of a resource, a resource whose scarcity is politically and economically constructed and cultivated. A managerial and planning system concerned with distributional issues is of vital importance. This would simultaneously question existing power relations in the city and inevitably raise the critical questions of 'Whose water?' and 'Whose city?' we are talking about. Challenging the systematically uneven power that characterizes both formal and informal water supply systems in the Latin American city requires a closer cooperation of the local people in the management of and control over nature and its water. Increased and improved citizen participation, and the reversal of the productivist bias could lead to a reorientation of water politics to one based on considerations of equal distribution and just accessibility. In particular, those who are dependent on private door-to-door water selling by water vendors pay the price for this uneven distribution of water. This is the theme we shall turn to next.

7

The Water Lords: Speculators in Water

7.1 The private local water economy

As already documented in Chapter 3, more than 600,000 of Guayaquil's inhabitants depend on the 'tanqueros' for their daily supply of water. Private water vending is of course not a recent phenomenon. It was a common activity in the time of the Incas, and became the standard means of urban water provision in the eighteenth and nineteenth centuries. With the introduction of urban water engineering systems in the late nineteenth century water distribution became increasingly organized by the state, while new engineering practices aimed to provide the entire city with access to water. However, with the exception of a few years after the opening of the La Lolita treatment station in 1928, Guayaquil never really succeeded in achieving the objective of full coverage. Nevertheless, the aim of providing unlimited quantities of potable water for all of the urban population at a marginal (highly subsidized) price was never abandoned, and has been built into successive Master Plans until this very day. The political economic realities of Guayaquil's urbanization process ran counter to this objective, for reasons discussed in previous chapters. Although the plans always held up the promise of unlimited and guaranteed water supply, a promise which served very important political and ideological functions as it deflected potential social unrest, cultivated clientelist political programmes, and contributed to legitimizing privatization, a growing number of people became systematically excluded from access to publicly provided water.

Particularly during the period 1960–90, there was a growing gap in water coverage. Whereas 73% of the urban population was connected to the public water system in 1974, this declined to just 64% in 1990. In absolute terms, 222,269 people were deprived of connections in 1974, but by 1990 this number had risen to 596,013 (according to conservative INEC data). According to the 1980 Master Plan, 75 to 80% of the metropolitan population was connected to the supply system in 1980, while only 20% was serviced by tank lorries (224,964 people). This means that there has been an almost threefold increase in the

Fig. 7.1. Turning H_2O into money.

number of people who are dependent on private water purchases in just over little more than a decade (EMAP-G 1980: Cuadro 4.4–16).

A growing army of 'tanqueros' serviced these people, who were invariably living in the new invasion settlements of Guasmo, Suburbio, Bastion Popular, Isla Trinitaria, and several other more recent invaded land settlements. According to the 1980 Master Plan, the average daily delivery of water by tanqueros amounted to 3,050 m^3/day (EPAP-G 1991*a*), approximately 2,700 m^3/day in the metropolitan area alone. A few years later (1982), the volume distributed by tanqueros had risen to 8,702 m^3/day (Universidad Catolica 1982: 1), and by 1992, this figure had increased further to an average of 12,000 m^3/day. Demand for water is considerably higher in the wet season (winter months), with a peak in January, and lowest in the dry season (running from May until December). The highest demand for tanquero water was in January with an average daily supply of 14,591 m^3/day, falling to about 8,371 m^3/day in July (data for 1992). Official data for two months in 1992 are shown in Table 7.2.

This suggests that the average daily per capita consumption of water in the areas supplied by tanqueros is 20 litres per day per person. Of course, these data are based on official statistics, whose accuracy is notoriously unreliable and usually underestimate the real situation. The immense fraud with and theft of water is not included in these data. According to EPAP-G estimates, only 37% of the water is accounted for and 17% is stolen (see Chapter 6).

Table 7.1. Number of 'tanqueros' active in the city of Guayaquil, 1992

Filling station	Number
Bellavista[a]	271 (+36[b])
Duran	34
Via a Daule 8½	41
Via a Daule 12½	33
Via a Daule 16½	41
TOTAL	420 (+36)

[a] The filling station of Bellavista was relocated to Via a la Costa km 10 in April 1993.
[b] These 36 belong to the Defensa Civil. They currently use (together with other tankers belonging to public services) the Chlorinadora at the Universidad Estatal as their filling station.

Source: EPAP-G.

Plate 7.1. 'Tanqueros' filling their water lorries at the filling station.

In 1990, there were a total of 383 tankers of which 302 supplied households, while 81 were registered as industrial water suppliers, whereas by 1992, 420 privately owned tankers operated at the five filling stations (see Table 7.1 and Plate 7.1). In addition, illegal vendors on tricycles sell water in smaller quantities (10 litre buckets sold for approximately 200 sucres in 1992 values) in areas

Table 7.2. Distribution of water by tankers in January and July 1992

	January	July
Volume water (m^3)	364,775	209,288
Volume water/day	14,591	8,371
Number of trips	47,781	25,538
Trips/day	1,911	1,021
Trips/tanker/day	4.55	2.43
Value (million sucres)[a]	25.53	15.09

[a] Exchange rates: Jan. 1992: 1US\$ = 1,323 sucres; July 1992: 1US\$ = 1,504 sucres.

Source: Data from registration lists of 'tanqueros' at EPAP-G.

with serious shortages. Some of the smaller towns on the Peninsula, such as Salinas, Libertad, and Playas, are supplied by 129 tankers that deliver water to households and an additional 53 suppliers of industrial water. A total of nine water boats supply the shrimp factories and customers on boats and islands (data from EPAP-G). In short, over 600 tankers are active in the area covered by EPAP-G's water supply region.

With an average daily delivery of 28.5 m^3/day and an average carrying capacity of 7.4 m^3/tanker, they each do on average four or five delivery trips each day (although daily delivery rounds can vary from as few as two or three to more than ten). In contrast, only about 102 tankers were active in the city in 1979—approximately one-sixth of the number operational 13 years later. This proliferation of private water sellers indicates the spiralling severity of the urban water crisis.

7.2 Selling water door to door: making money from monopolizing nature

The quasi-monopoly control held by these private water vendors over a key biological and social commodity allows for a massive concentration of social and economic power in their hands. This is most vividly illustrated by analysing the water rents that can be extracted by the water vendors as a result of their exclusive access to potable water. First of all, there is an enormous discrepancy between the water price charged by the public water authorities for those that can enjoy the luxury of domesticated water and the price charged by private water vendors.

Table 7.3 shows the multiples of price charged by private urban water vendors in comparison with the 'official price' charged by the public water util-

Table 7.3. Price multiples and water prices charged by water vendors, mid-1970s to 1980s and 2001 (unless otherwise indicated)

City	Country	Multiples of water charged by public water utility[a]	Water price[b] US$/m³ 1988
Data for mid-1970s to 1980s			
Port au Prince	Haiti	17–100	
Tegucigalpa	Honduras	16–34	
Lima[j]	Peru	17	20–50 intis
Barranquilla[f]	Colombia	28	2.00
Mexico City[g]	Mexico	40–114	400 pesos
Guayaquil	Ecuador	200–300	2.11–3.16
Quito[i]	Ecuador	27	1.70–2.00
Barquisimeto	Venezuela		1.00–1.40
Cochabamba	Bolivia		1.40
Data for 2001[e]			
Baranquilla	Colombia	10–12	5.50–6.40
Guatamala City	Guatamala	7–10	2.70–4.50
Lima	Peru	8–10	2.40
Guayaquil[h]	Ecuador		3.20
Cochabamba[c]	Bolivia	5	2.40
El Alto[d]	Bolivia	16	3.30

Sources: [a] World Bank (1989: Table 3.2, p.70); [b] Vásconez (1991: 51); [c] Crespo (2002*a*: 111 and 117); [d] Komives (2001); [e] Solo (2001: 2); [f] Barranquilla (date 1989) Bernal (1991: 154), [g] Mexico City (date 1983) Bataillon and Panabière (1988); [h] Guayaquil from field work (1993); [i] Quito's multiple (date 1987) Vásconez (1988*a*); [j] Lima's water price (date June 1987) Espinoza and Oliden (1988: 57).

ity as well as the price charged per cubic metre for a number of cities in the 'developing' world. In Guayaquil, private water vendors buy water from EPAP at 70 sucres/m³ and this is sold (in September 1993) for 4,000 to 6,000 sucres/m³ (800 to 1,200 sucres for a tank of 55 gallons or 200 litres). In the case of Guayaquil, the multiple can be up to 300 in 1993 if compared with the basic tariff charged by EPAP-G for households using less than 15 m³ of water each month. This alone indicates that poor urban residents have to pay up to 30,000% (*thirty thousand per cent*) more for their water (of an inferior quality) than the higher income residents living in urban sectors served by the public water system. In the context of a minimum wage (for those who have the luxury of being formally employed) of 60,000 sucres (US$30) per month, an average family of four with one income earner on minimum wage, using about 100 litres of water a day, would spend US$7.50 per month on purchasing water (or 25% of its disposable income). According to the Master Plan, in 1979 an average family of four would spend 173 sucres (US$6.22) per month on water purchased from tanqueros (EMAP 1980: cuadro 8.8). In 2002, the price to the consumer for a delivery of 200 litres had increased to US$0.80 (although this officially set price is often arbitrarily increased by the water vendors). For an

Plate 7.2. Selling water by the barrel.

average family of four, using on average 25 litres per person per day, the total monthly cost amounts to US$12.40 in 2002, a considerable cost for the majority of families living in the invasion settlements.

Table 7.4 shows the tariffs charged by EPAP-G. The basic tariff for domestic water usage (less than 15 m^3/month) remains fixed, while the increments are subject to a monthly increase of 3% for domestic uses and 4% for commercial and industrial uses. Even households that consume more than 300 m^3 of water/month pay considerably less (13 times less) than those who are dependent on the tanqueros. In addition, a fixed sum is added for maintenance. This value varies depending on the diameter of the supply pipe. The base rate is 70 sucres (for ½″) and moves up to respectively 100 sucres for ¾″ pipes, 300 sucres for 1″, and a maximum of 3,500 sucres for a pipe of 6″. Moreover, residential users pay an additional 78% of their water charge to cover sewerage costs. Commercial users pay an increment of 96% and industrial users 101%.

The increasingly acute water supply crisis gives the tanqueros a uniquely powerful economic position in the urban economy. They buy the water from EPAP-G at a highly subsidized price. Until 1987, the price per cubic meter paid by the water vendors was 8.4 sucres, a price that had remained unchanged since at least 1979 (EMAP 1980). In 1990, the price had risen to 15 sucres/m^3 for domestic water and 70 sucres/m^3 for industrial water. Because of fraudulent practices (selling water bought as domestic water to industry), the water price in the city was increased to 70 sucres/m^3 for all usages in October 1991. Despite

Table 7.4. EPAP-G tariff structure for residential water use, October 1988, December 1992, September 1993[a]

Value (sucre/m³/month)	Oct. 88	Dec. 91	Dec. 92	Sept. 93
Residential use				
Basic tariff[b]		300		
16–30	30	92	131	171.6
31–60	40	124	175	228.8
61–100	50	155	220	286.0
100–300	60	186	264	343.2
>300	70	214	307	400.3
Commercial use				
Basic tariff[b]	900		6,396	
16–50	70		498	
51–100	80		568	
101–300	90		639	
>300	100		711	
Industry average, Types A–C[c]				
Basic tariff[b]	1,350		9,593	
16–50	100		711	
51–100	110		781	
101–300	127		924	
>300	143		1,066	

[a] The monthly increments for domestic water use increase are 3% and 4% for commercial and industrial use.
[b] 0–15 m³/month.
[c] Each industry is placed in one of three categories on the basis of its water requirements.
Sources: EPAP-G (see also EMAP-G 1988).

attempts by the governor of the province to set maximum prices for a tank of 200 litres (55 gallons), the private water vendors are able to increase water prices arbitrarily. In 2000, an attempt was made to establish 'official' prices, although these remain high for the customer. In September 2002, the tanqueros were buying water from the water company at US$0.33 for 1,000 litres and selling it to the urban residents at US$0.80 for a 200-litre (55 gallons) tank, a gross profit rate of almost 90%.

Table 7.5 shows the recent evolution of the real water price asked by the water vendors expressed in current dollar terms. The prices listed are 'standard' prices, but these are frequently increased arbitrarily, for example under conditions of scarcity. Moreover, prices tend to go up in areas further away from the filling stations. For example, in Isla Trinitaria, prices tend to be 10 to 25% higher than elsewhere. A field survey in September 1993 indicated a price range varying from 800 sucres in Duran, 1,000 sucres for water from the Via a Daule filling station (for Bastion Popular and surrounding settlements) and 1,200 sucres for water coming from the new filling station located at Via a la Costa km 10. Over half of the tank lorries are filled up at this station. The price can go up

Table 7.5. Evolution of official and real water prices (for a tank of 55 gallons or 200 litres)

Date	Official price in sucres	Real price in sucres	Exchange rate sucres/US$	Real price in US$
1976	2.50		27.45	0.09
13/08/79		4.00/8.00	27.80	0.14/0.29
07/09/87		60.00	193.8	0.31
14/07/88	60	80.00	436.2	0.18
24/04/89		120.00	569.2	0.21
05/07/89		150.00	600.0	0.25
23/05/90	150	200.00	821.5	0.24
20/03/91		250.00	1,014.0	0.25
14/07/91	200	300.00	1,119.0	0.27
31/08/91	250			
06/06/92		400.00	1,477.0	0.27
11/09/92	400	700.00/800.00	1,828.0	0.38/0.44
11/11/92	450			
22/04/93	550	800.00/1,200.00	1,895.0	0.42/0.63
June 2002				0.80

Sources: Newspapers, field work, interviews, Plan Maestro.

to 1,500 sucres for 200 litres in Isla Trinitaria. By 2000, after the dollarization of the economy, the price was set at US$0.80.

The data also indicate that the real price for water (in dollar parity terms) has gone up, particularly during the past few years. The combination of galloping inflation, national government austerity measures, rising cost structures (notably for gasoline), and growing water scarcity allowed water vendors to raise their prices indiscriminately. By the end of 1993, the private water price had risen to between US$2 and US$3 for 1,000 litres and continued to rise afterwards. The difference between the purchase price of water (70 sucres for 1,000 litres) and the sale price (between 4,000 and 6,000 sucres for 1,000 litres) has increased from 230 sucres in October 1990 to 5,930 sucres in 1993 for the case of Guasmo, Suburbio, and Isla Trinitaria. This corresponds to an increase in dollar terms from US$0.20 to US$3.20. By 2002, the 'wholesale price' of water was US$0.33 per 1,000 litres, which would 'retail' at US$3.20, a gross margin of US$2.87 for each 1,000 litres sold.

7.3 The water power of the tanqueros

It goes without saying that the water vendors wield substantial economic power. As a result of their quasi-monopoly position in delivering water in the areas not covered by the 'official' network, they are able to extract sizeable

Table 7.6. Ownership structure of water-vending trucks in 1992

Number of trucks owned:	1	2	3	4	5	6	7	8
Number of cases:	134	60	15	14	4	1	0	2
Institutions: 23								

Note: These data exclude the trucks operated by the Defensa Civil.
Source: EPAP-G.

amounts of water rents from the poorest segments of the urban population. On the basis of an average price of US\$0.80 for 200 litres (which is considerably higher in some areas) and an average daily delivery of 12,000 m^3, the poorest communities pay daily US\$48,000 to the water vendors. On an annual basis (300 working days), the total annual cost of water borne by the 600,000 residents who do not have access to water amounts to US\$14.4 million. In 1993, the estimated total annual return to the private water-vending industry was US\$9.5 million, and by 2001 the annual income transfer from the most marginal segments of the urban population to the water vendors under the form of monopoly water rents had increased by more than 50%.

This means that the average gross annual income for each tanker truck rose from about US\$22,619 in 1993 to an estimated US\$32,285 in 2001, based on an estimated total of 420 tanqueros. At a more conservative estimate of 305 privately owned trucks (based on EPAP-data for October 1992), the gross turnover for each truck is US\$47,200. As suggested by the data in Table 7.6, some tanqueros own more than one tanker. Consequently, their annual turnover has to be multiplied by the number of tanker trucks they own. The data for this table were collected on the basis of the officially registered names of tanker owners at EPAP-G. Clearly, it may well be the case that real ownership patterns vary as there may be a discrepancy between the list of formal owners as provided by EPAP-G and the real (but hidden) ownership structure. According to these data, only 136 owners reported one tanker ownership, the others controlled more than one and 21 registrants owned more than three tankers.

In order to assess net profitability, we have to estimate the actual cost per 200-litre tank of water. Over the past few years, a number of organizations have attempted to estimate the real cost of selling water by means of tank lorries. In order to compare the various estimates, which were undertaken at various times during a period of hyperinflation, we shall use current dollar values in order to compare the various calculations. The most comparable data were available for a period during the early 1980s. Of course, average cost calculations depend largely on assumptions and estimates of average tanker capacity and average number of daily rounds. Table 7.7 shows the results of various

Table 7.7. Comparison of average cost per tank of 200 litres of water in current sucres and US$

	Source and date of calculations						
	Frente de Usuarios		FEDEBAS			Asociación Tanqueros	Defens Civil
	20/9/90	11/9/92	6/92	9/92	9/93	24/9/90	4/9/92
Average capacity	8	7	8	8	10	7.5	10
Number of rounds/day	9	10	8	8	8	8	5
Number of tanks sold	346	350	320	320	400	280	225
Vehicles (million sucres)	25	50	25	30	140*	35	—
Depreciation period (years)	10	10	5	5	8	10	—
Depreciation of vehicles (costs per day)	8,333	20,000	16,733	20,000	58,333	11,644	—
Operating costs per day:							
Vehicle maintenance	8,500	23,300	17,090	20,508	13,613	21,952	—
Wages	12,000	20,000	21,400	28,000	27,000	14,400	8,666
Gasoline	9,720	39,420	17,640	40,880	46,720	14,400	29,200
Other	500	—	—	—	13,000	500	—
Water cost	1,080	4,410	960	960	5,600	900	750
Cost of 200-litre tank:							
In nominal sucres	115.99	287.00	227.00	344.80	410.60	227.62	
129.36							
In US$	0.14	0.16	0.16	0.19	0.22	0.28	0.(
Exchange rate, sucre/US$	821.91	1,828.00	1,477.00	1,828.00	1,895.00	821.91	1,828.(

* New Vehicles.

Sources: Asociación de Tanqueros '25 de Octubre' (1990); Federación de Barrios Suburbanos de Guayaquil (FEDEBAS-((1992*a*,*b*; 1993); *Hoy*, 22 April 1993; Frente de Usuarios del Guasmo (1990; 1992); Comité Ejecutivo Pro Agua Potable pa el Suburbio Oeste (1989).

calculations. The source of these calculations varies from the Asociación de Tanqueros', the 'Defensa Civil', the 'Frente de Usuarios' to the 'Federación de Barrios Suburbanos de Guayaquil (FEDEBAS-G)'. Clearly, each of these groups represents different interests and objectives and, consequently, their estimates and assumptions vary accordingly.

The estimates of the Defensa Civil exclude investment costs and are, therefore, the least useful. The other cost estimates vary from US$0.14 to US$0.28 per 200-litre tank. The median cost calculation is US$0.21. Our own estimate, based on interviews and cost calculations, also points at this average as a reasonably accurate estimate of the cost structure of private water vending. On the basis of this, we can calculate total profits and rent extraction generated from the private appropriation and distribution of nature's water.

The average daily distribution of 12,000 m^3 water gives a total cost of US$12,600/day or US$3,780,000/year (7.16 billion sucres at 1993 rates). The annual net profit from water vending can be calculated at US$5.72 million. For each individual tank lorry, the annual cost is US$12,393, which when related to an annual turnover of US$31,148 gives an average annual profit of about US$18,755 (35.54 million sucres) for each of the tanqueros (data for 1993). For those who own more that one tank lorry, this average annual profit has to be multiplied by the number of tankers owned. These water profits compare with a minimum national wage of 60,000 sucres/month (in 1993) or, if we include earnings from the informal economy, an average monthly family income of about 150,000 sucres (of 1993) in the marginal areas of the Guayaquil (based on interview and questionnaire data).

This transfer of value, extracted from the poorest communities, inevitably ends up in dollar accounts in Guayaquil's or Miami's banks. It is a suggestive analysis of how nature's water is socially transformed into the power of money, which, through water rent extraction, is transferred from the poor residents to the water speculators. This capital, then, often disappears from the economy through the transfer of sucres into dollars as a shield against inflation (before the dollarization of the economy). Ironically, the Miami dollar-holding accounts, in turn, often find their way back into the economy by means of international lending mechanisms. This circulation of money and water tells the story of how these loans are themselves financed through the appropriation of Ecuador's nature, the interest on which has to be matched by a further extraction and transformation of nature's resources.

Indeed, this analysis shows how the appropriation of nature's water implies a socially highly stratified mechanism of access to nature. Nature's water is hereby turned not only into stuff for satisfying human thirst but also into a vehicle for exploitation and the generation of social power. The urbanization and urban circulation of water embody a series of mechanisms that generate, consolidate, and reaffirm mechanisms of marginalization, exclusion, and domination. While the marginal settlements grew out of speculative developments and the occupation of marginal lands sold at very low prices for clientelist political reasons (10 sucres/m^2), the actual costs of occupying this land are multiplied many times through the extraction of extortionate water rents by water vendors. Moreover, the growing social scarcity of water solidifies the power of water and intensifies the sway of the water owners over the life and health of the urban population.

7.4 Private water vending/state control: an ambiguous symbiosis

Until early 2001, Guayaquil's urban water supply system was characterized by an ambiguous public/private relationship. While the transformation of nature's water was a wholly public activity during the twentieth century, the urbanization of water was structured through a combination of a centralized

Table 7.8. The average income and expenditure structure of EPAP-G's water production

	1979	1982	1990	1991
Source	Master Plan	Universidad Catolica	EPAP	EPAP Finance Department
Running expenditures (US$m)	9.6		13.6	16.94
per m³ produced	0.12		0.08	0.096
per m³ sold	0.20		0.20	0.26
Income:				
Water sales (US$m)		9.64		12.35
Other (US$m)		2.28		0.56
Per m³ sold		0.21		0.20
Transfers:				
Loans		9.7		
IETEL or EMETEL		1.54		2.66
FONAP		0.52		
Emergency contributions				32.13
Other		1.3		1.26
TOTAL INCOME		22.9		49.2
Total cost/m³ water:				
produced (in US$/m³)		0.27		0.28
sold (in US$/m³)		0.36		0.75

public system and a decentralized private system. Ironically, the public system supplied the more powerful industrial and social strata, while the poor were serviced through a private system of quasi-formal or informal water vending. Consequently, an ambiguous relationship developed between public production and supply on the one hand and private distribution on the other. In order to assess this relationship, we first have to turn to the financial and budgetary situation of EPAP-G, the recently privatized water supply company. The process of privatizing this company displaces these ambiguities, while perpetuating a condition of significant inequality in access.

7.4.1 Public water sales: subsidizing the rich

In order to assess the real cost of producing a cubic metre of potable water, we have to distinguish between operating costs and investment costs. EPAP-G was an organization that ran significant deficits, and whose running costs massively outweighed the income generated from water sales (see also Faudry 1997). For the moment, we shall exclude investment costs and concentrate on running costs. Table 7.8 summarizes average running water production costs at various times. Data are from various sources as well. In the first half of the 1980s (before the upgrading of 'La Toma'), the average cost of producing a cubic metre of water varied between US\$0.12 and US\$0.14. After the upgrading of 'La Toma' to an average daily production capability of 480,000 m^3, the average running costs fell to between US\$0.08 and US\$0.096/m^3. However, as a result of the enormous physical and commercial losses, the actual average cost of a cubic metre of water was considerably higher and fluctuated between US\$0.20 (in 1979 when the average loss amounted to 40%) and US\$0.26 in 1991 (when losses had risen to 63% of total production). With the exception of high-volume consumers, the price paid by the consumers was far below the actual cost of producing the water. In 1991, the income generated by EPAP-G from selling water was on average US\$0.20 per litre, while the cost was US\$0.26.

If we add investment costs and the repayment of interests and loans to this, the actual deficit was quite considerable. This suggests that the real cost of producing and distributing water was considerably higher than the revenue generated from water sales. The difference had to be made up through a variety of taxes, which, in the end, have to be paid by the whole of the population. These included a variety of local and regional taxes, a 10% contribution from the Telephone Company IETEL, and emergency loans from the national government. The government also had the responsibility of servicing (part of) EPAP-G's debt. Given the highly exclusionary system of water distribution, it is ultimately the elites and the middle classes who received most benefits as they can enjoy highly subsidized water at a fraction of the real cost of producing and distributing it. This system of cross-subsidization ultimately has perverse social effects (Yepes 1999).

These calculations exclude the recovery of the investment costs associated with expanding and/or upgrading the system. During the early 1990s, two big projects were implemented. The first one, financed through a US\$72 million loan from the Spanish government, was the construction of a new water treatment station next to the existing one at 'La Toma', which (at least theoretically) doubled the production capacity to almost 1.5 million m^3/day. The other project concerned a US\$51.5 million loan from the World Bank to upgrade and expand the water distribution system. The latter loan was delayed several times due to EPAP-G failing to comply with World Bank conditions associated with increased efficiency, improved water accounting and the fusion of the sewerage and water distribution companies (which was eventually completed in mid-1993). The Inter-American Development Bank finally granted a loan (see Chapter 6) to facilitate the transition to a privately organized water and sewerage system. In the end, of course, the public pays for the cost of privatization, as these loans have to be serviced by the state. Given the high debt burden of the Ecuadorean state, these loans accentuate an already precarious financial position.

In addition to these loans, EPAP-G's programme of investments until the end of 1994 amounted to a total of US\$122.6 million, as well as 69.96 billion sucres (of 1992) equivalent to a further US\$47.2 million. At the time, it was estimated that a total investment requirement of over US$170 million was needed just to service the most urgent needs (EPAP-G 1992*a*). Clearly, Ecuador's problematic position in the international financial system rendered it quite impossible to raise such investment capital on the international financial markets. It was already evident that the majority, if not all, of these planned investments needed to be shelved indefinitely. However, these plans served important ideological and political functions by suggesting that the government is indeed taking charge of the problems and that solutions are imminent. The 'only' stumbling block is the international community, which is not sufficiently forthcoming with aid or loans! Eventually, the conundrum faced by the state also helped to pave the way for privatization. The private sector and the privatization of water supply became staged as the panacea that would finally solve the enduring problems of the city's inadequate urbanization of water.

7.4.2 Maintaining the status quo: the ideology of underdevelopment

Of course, the public water authority's deficits resulted in perpetual dependence on external financial sources to cover shortfalls in running costs as well as to finance new projects. This precarious financial position led to a deteriorating water supply system and the perpetuation of exclusionary practices. As discussed in Chapter 6, the emphasis on producing water led to a preoccupation with ambitious production facilities while management, maintenance, and distribution remained relatively neglected. The emphasis on major projects served as an ideological smokescreen that enabled lip service to be paid to a developmental discourse while putting the blame for the problems on external factors. The new water treatment station at 'La Toma' is a good illustration of this (EPAP-G 1991*b*). The absence of distribution networks, combined with the enormous physical and commercial losses, led to an only marginal improvement of water accessibility for those who were hitherto excluded from access to potable water. While the discourse staged by EPAP-G maintained that the new plant would provide a panacea for the city's water problems, the population in the marginal settlements would remain excluded unless a drastic reorganization of the distribution system were to take place.

Ironically, the tanqueros applaud these developments, as they represent the opportunity for them to maintain their water monopoly in the marginal settlements. The preoccupation with producing water maintains the power position of the tanqueros in terms of their control over distribution in the marginal settlements. Yet, at the same time, the income generated from distributing water in these settlements would be sufficient to finance all necessary water infrastructure works. In 1993, the estimated annual payment for water from the 600,000 excluded inhabitants amounted to US$9.5 million (18 billion sucres) for an average daily distribution of 12,000 m^3 water, an amount which doubled during the subsequent decade. This shows that residents, if threatened with death from thirst, show a very high willingness to pay for water. In 1991, EPAP's total income from selling water (excluding payments made for installations) was almost 14 billion sucres, or US$12.35 million (for an average daily distribution of 440,000 m^3 (97% of the available water)). If the water utility were capable of capturing the income now appropriated by the water vendors, self-financing would be a highly realistic prospect.

In sum, the investment and project policy pursued by EPAP-G perpetuated the chronic dependence of the marginal settlements on private supply systems, whilst the continued structural exclusion leads to a massive drain of resources from the marginal communities to the private water vendors. No wonder, therefore, that the water vendors support EPAP-G's policies as well as the privatization process, as they are the best guarantee that the private vending monopoly will continue in the future. This chain of processes locks the urban poor into a condition of structural underdevelopment and results in an uneasy alliance between the political elites on the one hand and the private water monopoly holders on the other.

7.5 Privatizing the contradictions of urban water supply

The proponents of the privatization process promise an end to the practice of lorry-based door-to-door vending of water and a more efficient and effective water delivery. However, the contradictions of urban water supply in Guayaquil remain as acute as ever, with the supply of water becoming part of the investment decisions and strategic considerations of a global utilities company. In January 2001, a private 30-year concession was awarded that became operational in April 2002 (see Chapter 6). As documented in the previous chapter, the contractual conditions of the new privatized management system look promising, yet the practice of private water vending has not disappeared. At the same time, the profitability of the private company depends on the twin conditions of increasing water rates on the one hand (the permission for rate increases is performance-related) and the successful improvement in the payment of water bills (which remains a formidable obstacle). While the state can now transfer blame for inadequate water supply onto the private company, the inequalities of access to urban potable water are likely to remain for the foreseeable future and the private water vending economy will continue to thrive.

Under these conditions, in which a significant proportion of the urban population is engaged in a difficult daily quest for water, it is no surprise that intense urban social struggles develop over water availability and distribution. It is these struggles that we shall turn to in our next chapter.

8

Contested Waters: Rituals of Resistance and Water Activism

'Whiskey's for drinking, water's for fighting about.'

(Mark Twain)

The discussion in the previous chapters has shown how water is deeply embedded in the practice of everyday urban life, and how the uneven power over its control and the oppression caused by its absence result in it being highly contested terrain. A wide spectrum of social conflicts and struggles consequently unfold over the appropriation of and access to 'metabolized' urban water; struggles that are embedded in and embody social, political, and economic power relations. Put simply, the transformation of nature and the urbanization of nature's water express the political economic and socio-ecological power relations that shape the urbanization process itself. The urban hydrosocial cycle is indeed infused with a myriad of social tensions and is contested terrain at each moment of its flow. We shall consider not only the strategies of the weak and the weapons they deploy, but also the tactics of the water vendors. Their control over water enables them to mobilize a range of tactics to maintain if not expand their hold over water and the appropriation of water rents. This chapter will explore these multifaceted dimensions of the social struggles around water and their political, economic and social significance.

8.1 Urban social struggles and the citizen

The most striking, if least powerful, actions are those waged by urban communities to gain or improve access to water. From the position of those facing oppressive mechanisms of water supply, four strategies have been identified (see, among others, Espinoza and Oliden 1988): passive acceptance, individual resistance, self-help, and social protest/mobilization.

8.1.1 Passive acceptance

Notwithstanding the immense problems associated with problematic access to water, acceptance of the exclusive and/or exploitative status quo is very often the most common behaviour, resulting in the absence of collective action. Cotic and Dascal (1987) suggest that the nature of an eventual possible, but essentially private solution (i.e. domestic connection to a comprehensive networked system), results in passive acceptance as a particular form of response. While the demand for the provision of roads, schools, health services, or public transport is often subject to collective action and struggle, water (and waste disposal (Olaya 1991)) has rarely resulted in popular and collective revolts. In the case of Guayaquil, water shortages are at the top of the agenda of many residents of the invasion settlements, although their precarious economic and political condition renders them powerless in the face of the perceived dimensions of the problem. The overwhelming majority of the urban population in these settlements is faced with the everyday struggle for survival that leaves very little time, money and/or energy for collective and organized struggle.

Despite these constraints, water issues do become important arenas for urban social struggle. A whole host of power strategies unfolds around water and structures the rituals of everyday life as well as wider urban and national political economic processes.

8.1.2 Individual resistance

Many forms of active, but basically individual, resistance shape the day-to-day struggle for water. These actions vary from the construction of illegal connections, the installation of water pumps to suck water from the underpressurized mains, and theft of water, to arguments with water vendors or the water utilities and, above all, playing on the intricate but deeply entrenched clientelist networks through which political affiliation, power, and control are established, exercised, and maintained. Such strategies are clearly divisive, inherently conservative, and feed an individualized, fragmented, and divisive urban political economy based on personal relations, favours, and rewards. In Guayaquil, illegal water connections and water theft are part of the everyday practices used to gain access to the precious liquid. Numerous newspaper reports suggest that intricate mechanisms exist to evade regulations and, in some instances, sheer necessity demands illegal actions of water appropriation. The installation of pumps to suck water out of mains is a common occurrence. Clearly, extracting water in this manner increases the danger of contamination and diminishes water availability downstream of the connection. In addition, breaking pipes contributes to higher maintenance costs and an increase in the percentage of water lost through leakage and spillage.

Of course, the powerful have sufficient resources to dig their own wells to tap underground water. Top hotels and key industries construct their own water supply system, usually by digging wells and tapping into the aquifer. However, the unregulated tapping of underground water has unknown effects on the water table and subterranean water dynamics. In addition, water quality from underground aquifers is of varying quality as pollutants of all kinds have begun to contaminate groundwater sources.

8.1.3 Self-construction and self-financing

The response to inadequate services and facilities can take the form of cooperative self-organization. This is often aided and actively supported by the state or quasi- or non-governmental organizations, and is an example of people claiming their right to the city and to nature in an active, collective, and organized fashion. Although almost inevitably riddled with conflict along the private/collective interest on the one hand and along the state/communitarian cleavage on the other, self-construction does tend to undermine clientelist and paternalist politics (see Montano and Coing 1991 for Buenos Aires), helps to define the problems at a wider social level, and recognizes that the heart of the problem revolves around wider issues of social reproduction and collective emancipation (Craske 1993). These forms of self-help may take the form of collective but illegal community-based water tapping and conduction (see Ridgley 1989), or may be based on actively soliciting the support of the state (as in the case of some of the barrios of Quito (see Comite Pro-Agua Potable y Progreso Comunitario 1990)) or of international aid organizations (as in the case of Cusco (Caballero 1991)). In Guayaquil, local communities, sometimes in close cooperation with the water authority, have installed community water taps, which are connected to the mains and are managed on the basis of a decentralized community system. In 1992, for example, EPAP-G had 109 registered community taps installed which each serviced a number of housing blocks in housing cooperatives.

In another case, FODUR (Fondación de Desarollo Urbano), a state-backed community development organization, has constructed a series of 'Cisternas Comunitarias' (collective cisterns) in the marginal settlements. This is often done in close cooperation with local communities, with funding organized through the state (which provides 80% of FODUR's funding) and supplemented with income from interest on savings accounts at the Banco de la Vivienda. These projects are implemented with the help of local people, and once completed, daily project management is transferred to the water company. These 'Cisternas Comunitarias' feed a series of up to 40 taps, each of which services approximately 100 families. The cistern in Prosperina and the two cisterns in Mapasingue are fed with water pumped uphill from the mains, while those in Bastion Popular are fed directly from the mains (interview with C. Vaieharle, Director of FODUR, 20 Oct. 1992). In Guasmo, 47 smaller 'Cisternas Comunitarias' operate, but as there are no mains in Guasmo these have to be filled by tanqueros. Although the price for bulk water delivery is much lower (*c*.20,000 sucres (in the autumn of 1992) for a load of *c*.7.5 m^3) than for individual delivery, tanker lorry owners are not eager to fill the cisterns, and only serve cisterns after the more profitable deliveries to individual households are completed. Consequently, their success is rather ambiguous and some of the cisterns have never really functioned properly. For example, on 11 November 1991, *El Universo* reported that only two of the 47 cisterns operate normally. FODUR also demanded funding for the purchase of 100 tanker lorries to break the monopoly of the private water distribution system, although this funding has never been forthcoming. In the mid-1990s, an international NGO began to undermine the monopoly of water vendors by introducing community-managed water trucks (see Chapter 9).

8.1.4 Urban social struggle and organized grass-roots mobilization

Organized forms of urban social action usually seek a degree of structural transformation, which '[t]he clientelist system could never produce not only because they [the demands] are too universal (not aimed at rewarding specific blocks of voters), but also because they seek to formalize the relationship between the State and its citizens, an event that would threaten the very basis of power enjoyed by political patrons' (Rogers 1992*b*: 19). In addition to pressuring the state and its organizations to do something, struggle unfolds around the exploitative and oppressive mechanisms associated with the water vendors' exclusive control over water. These forms of social struggle in the sphere of reproduction can take a variety of strategies such as mass demonstrations and action (as seen in Lima) (Zolezzi and Calderon 1985; Espinoza and Oliden 1988), payment strikes, occupations of offices, road blockages, taking water officials or water vendors as temporary 'hostages', and preventing 'normal' task performance (see, for example, Benaiges 1991; Bennett 1988; 1995). In some cases, like in Cochabamba or Brazil, they can take the form of collective and generalized protest (Crespo 2002*a*; de Oliviera Filho 2002).

These struggles can arise spontaneously as wildcat actions, particularly under conditions of extreme water scarcity. For example, water-meter operatives are occasionally molested during protests against water scarcity. In times of great shortages as a result of equipment and infrastructure breakdowns or strikes by tanqueros (see below), residents organize marches to the headquarters of the water company or to water filling stations to demand water and to exercise their 'right to nature'. Also, water payment strikes are organized to indicate discontent with the problematic water supply situation (*El Telegrafo* 4 Dec. 1989). Spontaneous rebellions and outbreaks of violence against tanqueros (often spilling over into full-scale rioting) do indicate the sensitive nature of the problem. For example, after the implementation of the austerity programme by the Sixto government in September 1992, which resulted in the average price of basic commodities rising by 50% (160% for gasoline and 100% for privately sold water), large-scale rioting started in a number of Guayaquileño neighbourhoods such as Bastion Popular and Flor de Bastion. Tanker lorry drivers were attacked, molested, and robbed, and had to flee while attempts were made to set fire to their lorries. The army had to be called in to control the rebellion, more than 30 people were arrested, and the area was militarized for a number of days until calm was restored (*El Universo* 15 Sept. 1992) (see also below). After the relocation of one of the filling stations from Bellavista to the Via a la Costa in May 1993, tanquero owners went on strike again, resulting in protest marches from residents to EPAP's headquarters. The first week of June 1993 was characterized by water-related protests, including road blockages, tyre burning, and demonstrations (*El Télégrafo* 2, 3, and 7 June 1993; *El Universo* 3 June 1993). During a long period of intense water shortages in 1998, marches were organized, protests staged, and skirmishes broke out at water filling stations. Mass demonstrations and gatherings of people demanding water at the filling stations are indeed regular occurrences in the city, just one of many indicators of the intensity of the social struggle that unfolds over the control of urban water.

In addition, a number of more or less well-established and organized community organizations and political movements have tried to mobilize popular movements around the water issue. These organizations, of which the Federación de Barrios Suburbanos (FEDEBAS) is the best known in Guayaquil, seek to unite various smaller neighbourhood-based groups in an effort to consolidate power and more effectively influence the state's urban policies (FEDEBAS 1996). They aspire to more universal goals for their demands; not only a more just distribution of goods and services but also inclusion in the power structure that determines the pattern of such distribution (Rogers 1992*b*: 7). Although it is often difficult for these organizations to escape links to particular political parties, and enduring systems of political patronage and clientelism tend to limit their effectiveness as representative institutions, they are quite effective as tools of political incorporation. Established in 1948, FEDEBAS is one of the oldest organizations, and is an urban social phenomenon dating back to the formation of Suburbio, the oldest invasion settlement which is now formalized. The president of FEDEBAS, Luis Gomez, admits that in the course of its history, the Federation 'has not escaped the pattern of political control' and the problem of political incorporation. Currently, the aim is to direct the struggle in a united and consciously non-partisan manner, whilst maintaining a deeply committed social position. The demands made by FEDEBAS address the need to redefine the relations between the state and the residents. Instead of the informal patron–client relationship of the past with its clientelist emphasis on the local provision of goods and services by the state, FEDEBAS seeks to institutionalize the duty of the state, not only to provide quality infrastructure to the residents of the city, but also to include representatives in the decision-making process leading to the allocation of urban services. For more than ten years, the area of water provision has been one of many in which FEDEBAS has been active (interview L. Gomez 13 August 1993).

Their water-related activities and struggles revolve around a series of issues, firstly involving direct negotiation with the water company. Five main demands are made in this context, which attest to the particular social and political objectives of this form of grass-roots resistance and struggle (FEDEBAS 1993):

1. To set up a community-based management commission including representatives from the water company, FEDEBAS, Defensa Civil, the military, and owners of tanker lorries. One of the tasks of the committee would be to study and implement an official water price.
2. To plan and implement a more efficient system of water distribution in the suburban areas.
3. To improve the supply of water for the 'Cisternas Comunitarias'.
4. To aim for a better sanitary control of the distributed water.
5. To set up a system to provide cheaper water by means of tanker lorries.

In addition, FEDEBAS lobbies for further infrastructure projects that would benefit the suburban areas, in particular Guasmo and Suburbio. Currently, they advocate the use of abandoned swimming pools as mini water reservoirs. For several years they have also advocated the initiation of a US$44 million project to use subterranean water south of the city (the El Chobo project—see Plate 8.1) (see Watkins 1991; EPAP-G 1992*b*), which would benefit the southern parts of the town, Duran, and part of the hinterland.

They also claim that half of the fleet of 36 tanker lorries owned by the Defensa Civil are used to supply hotels and industries. In August 1993, they charged these consumers 18,500 sucres for 7 m^3, compared to 17,500 sucres for the same volume for domestic water use or to fill 'Cisternas Comunitarias'. Finally, FEDEBAS attempts to undermine the private monopoly of the tanqueros by initiating self-help projects to take community control over water distribution. They work on a project aimed at acquiring a number of tanker lorries that would be managed and controlled by the community itself (see Chapter 9), while maintaining pressure on EPAP-G to take greater control over water distribution in the marginal settlements.

In addition to these project-based activities, FEDEBAS actively mobilizes local residents to protest against water scarcity (see Plate 8.1). They have organized a number of marches through the centre of the city, such as 'La Marcha del Balde sin Agua' in 1987 and again in 1989, and have continued to stage active protests throughout the 1990s. These marches represent a moment of solidarity and mobilization, a chance for people to affirm their identity and their struggle and to express their right to the city's nature. Of course, the relationship between the state and FEDEBAS is always tenuous, treading a fine line between the need to negotiate and to be partially co-opted on the one hand and the need to maintain an independent, mobilizing, and forceful collective voice and 'muscle-flexing' social stance on the other.

Other groups, such as the Comité Pro Agua Potable, loosely associated with FEDEBAS, were set up to mobilize and organize around the water issue. They organized a march to the National Congress in Quito in 1988 (*El Télégrafo* 4

Oct. 1988). Marches of this kind to various organizations and state institutions or to the filling stations to protest against the speculative practices of the tanqueros are quite a common occurrence. In addition, some pressure organizations, often more closely associated with political parties, have also taken a lead role in the struggle for water. For example, the Frente de Usuarios, loosely associated with the Partido Socialista Ecuatoriano (PSE), has been active in the water arena. The PSE itself has also tried to organize popular mobilization, such as a 'Marcha contra el Tanquero Ladrón' in 1990 (see Plate 8.1). In 1989, militants of the Movimiento Popular Democratico occupied the offices of the water company to protest against official price rises of water (*Expreso* 7 Dec. 1989). They clearly associated the water problem with five decades of clientelist based politics in Guayaquil and the resulting urban institutional chaos. The public water utility was considered to be a 'botin politico' for whoever is mayor at the time. Marches and organized protests tend to become more accentuated at important political moments, such as during mayoral elections, periods of major policy reform, or, of course, periods of scarcity.

These social struggles around water are, of course, deeply gendered (Bennett 1995; Moser 1987; Radcliffe and Westwood 1993; Stephen 1992). This is clearly associated with the detailed gender and labour division of water-related activities in addition to the gendered symbolic meaning (and powers) of water. Almost without exception, women take a central place in the struggle for better access to and control over water (Crow 2002). Bennett (1988: 19–20; 1995), for example, in her study of women and class struggle around water in Monterrey, Mexico, shows that women were the primary participants in two-thirds of the incidents for better water services. Women as the main users of water in the domestic sphere looked for strategies within their neighbourhoods, including blocking the streets with tubs, barrels, and their bodies. In addition to paralysing traffic and commercial and industrial activities they also engaged in actions affecting the productivity of the water authorities and private water vendors by seizing personnel and immobilizing trucks. In doing so, they explicitly challenge the unequal power over water by attacking water owners and controllers and expressing the deeply gendered relationships through which the urbanization of nature's water takes place. In Guayaquil, all the private water vendors are male without a single exception, and men equally take the overwhelming majority of key positions in the public water authority. There is consequently a virtually total masculine control over water supply and distribution, with the unsurprising consequence that women take a central place in the struggle for water. During occupations, street revolts, and other collective expressions of discontent, women play a pivotal role in leading and organizing the movement.

All these grass-roots mobilizations and actions raise the issue as to who has the right to water. Whose water is it? And for whose city! The right to water is directly related to the right to the city and to the meaning and practice of being an urban citizen. Yet these grass-roots mobilizations and forms of protest have

Plate 8.1. Collective social actions around water.

to be understood against the backdrop of the urban political institutions and the choreography of political power.

8.2 Contested water politics

8.2.1 Water struggles and clientelist politics

The above forms of grass-roots resistance to water deprivation are frequently highly visible and spectacular. This, however, does not make them any more effective in the face of the tactics used by those who hold effective political and/or economic control over water production and distribution. Indeed, the class tactics and strategies of the state apparatus, the private water vendors, the water utility company, and upper-class urban residents point towards an outright social conflict over the control of a key resource and commodity.

The history of clientelist and paternalistic political relations in Guayaquil prevents these class issues from being turned into class politics by particularizing and individualizing solutions that are granted in a very piecemeal fashion. This, in turn, ensures that the water-deprived citizens will be continually dependent on clientelist strategies. For example, when Elsa Bucaram was elected as Mayor of Guayaquil in 1988, she bought water and distributed it freely in a number of neighbourhoods in Guasmo and Suburbio to celebrate her election victory (*El Telegrafo* 5 May 1988). This is a classic example of clientelist politics, summed up eloquently by the same Elsa Bucaram: 'The people always respond, because it is easy to gain the hearts of the people. . . . It is sufficient to offer them (public) works' (Villavicencio 1988: 11) (my translation).[1]

Despite this populist gesture, water distribution deteriorated massively and rapidly during her term in the mayoral office, as did the organizational and administrative structure of the then municipal water authority. Of course, the acute water problem in the city was associated with five decades of almost uninterrupted clientelist politics in the city. The public water utility was considered and used as a 'botin politico' for mayor and councillors that allowed the partial satisfaction of popular demands and could be used effectively as a system of vote-bonding through providing services and jobs. During the 1980s and 1990s, for example, the payroll of the water company totalled more than 1,500 people, many of whom cashed in their monthly cheques without effectively performing any work (the pipones) (*Expreso* 8 Dec. 1989). It was not until there was a change of mayor in 1992 that hundreds of these 'pipones' were deleted from the municipal and EPAP's list of employees. Yet, the practice surfaced again a few months later. The transfer of authority over the water utility from a municipal to a provincial institution in 1989 was a not entirely success-

[1] El pueblo siempre nos responde, porque el corazón del pueblo es fácil ganárselo . . . basta que Usted le haga obras.

ful attempt at hollowing out the clientelist urban politics that dominated every realm of urban institutional life. This transfer, organized by the national state, took place in the context of a bitter and prolonged struggle between the populist mayor (Elsa Bucaram) who saw a key instrument of her clientelist politics being taken away by a left of centre national government. The conflict, which lasted for weeks and included the occupation of the water authority by municipal politicians, left the newly formed EPAP-G with a completely stripped-down infrastructure and a disorganized administrative structure. A similar reorganization of labour relations took place at the moment of the privatization of the water services (see Chapter 6).

Indeed, partisan and clientelist arrangements are deeply embedded in the choreography of political rituals in the invasion settlements. To a large extent, the development of these was part and parcel of the expansion and consolidation of clientelist political affiliations (Moser 1987; Scheers 2002). Either invasion organizers are politically affiliated or barrios (neighbourhoods) become co-opted through their local 'dirigentes' (barrio leaders) into a particular party allegiance. This allegiance is maintained and reinforced through the piecemeal granting of infrastructural and other improvements. Through the control of the local state, barrio groups that remain incorporated in the clientelist network are rewarded while those that make more radical demands remain unrecognized (Unda 1986). The dirigentes or local leaders play the role of intermediaries in the power brokerage, relying heavily on their political contacts as well as the informal networks they maintain in the settlements. They often exercise full control over land invasions and service provision—albeit often informally organized—in the barrio in exchange for political support and financial rewards. Alternative community groups are quite often suppressed or even violently silenced (Rogers 1992*a*). This clientelist patronage structure leads to a very fragmented, divisive, and particularized system of community action, and is an important stumbling block preventing the growth of citywide coalitions that could confront the state much more directly.

8.2.2 Water politics and exclusion

In Chapter 6, we discussed the social implications of the productivist logic of water utilities and the implicit class character of a systematic concentration of production that excludes or marginalizes the crucial issue of distribution. Indeed, the technocentric orientation of the public water authority whose strategies and investment policies target water production in the first instance, followed by a gradual extension of the water distribution network, assures a steady and more or less reliable flow of water to the urban elites. Maintenance and, in particular, strategizing around a more just distribution system seems to be one of the low priority areas. In this way, the 'urban water managers' (Pahl 1966) safeguard their exclusive control and allow the upper classes and their economic support structures (commerce and industry) to withdraw as much water as they desire at a highly subsidized price. The clientelist political network, in turn, prevents these class issues from being turned into class politics by particularizing and individualizing solutions, granting piecemeal improvements, displaying grand schemes and plans promising improvements, and blaming the capitalist world order for failing to achieve the necessary investments.

These water politics serve to keep the water problem high on the agenda and help to channel potential discontent into a technocratic discourse that focuses on engineering problems and solutions. In addition, portraying relative water scarcity as the result of deficient production capacity allows the transfer of blame to the 'Other', in particular the unwillingness of international lending bodies to provide the necessary financial backing to implement the proposed projects. Such strategies, although not entirely wrong-footed, help to alleviate social pressure from the local state and the public water authority and allow the existing water politics to continue without fundamental challenge. However, through mobilizing particular discursive strategies, these forms of 'institutionalized social exclusion' are always mitigated by the threats of social revolts, civil disobedience, and potential changes in the political landscape. This is readily recognized by the elite Colegio de Ingenieros Civiles del Guayas:

> The deterioration of the quality of life and of general well-being, to continue the current tendencies, could, in not too distant a future, lead to explosive levels of social resentment and to the despair of the affected masses and to unexpected social reactions that, once unleashed, become uncontrollable because of the ensuing chaos and ardour of the actions. (Castillo 1988: 4) (my translation)[2]

Clearly, such considerations, combined with the need to maintain clientelist ties, led to a situation in which EPAP needed to tread a fine line between ensuring the continued support of the political and economic elites by providing a guaranteed service on the one hand, and extending water supply to the invasion settlements to diffuse potential unrest and perpetuate the benefits of the patronage system on the other.

The productivist logic is further fed by the strategies and objectives of international lending agencies. Bilateral loans and World Bank assistance tend to focus on increasing production capacity rather than improving managerial or distribution practices. The conditions of international lending, such as requiring technology and expertise to be sourced from the donor country, further reinforce the technocratic productivist logic. In addition, the problematic foreign debt position of Ecuador and the related structural adjustment policies pursued since the election of the conservative Sixto government render uncom-

[2] El deterioro de la calidad de la vida y del bienestar general, de continuar las tendencias actuales, puede llevar en un no lejano futuro a niveles explosivos de resentimiento social y de la desesperaciòn de las masas efectadas y a imprevisibles reacciones sociales que una vez desencadena resulta incontrolable por el caos y la dureza de los enfrentamientos que desata.

mitted global money much more difficult to come by (Suarez-Torres *et al.* 1997). These tensions, of course, contributed to facilitating the political project of privatizing the city's water supply system.

8.3 Water terrorism: speculating with water

The tanqueros not only maintain a quasi-monopoly over water distribution in the invasion settlements, but also control the most direct and oppressive forms of socio-political action in the domain of water supply—the power of water speculation. Water supply and rent extraction can be manipulated by speculating with water during periods of water supply problems; withholding large sections of the population from access to water in order to 'manufacture scarcity'; or expanding the coverage of the non-serviced areas by sabotaging part of the existing water network.

8.3.1 Speculating on water shortages

As documented before, the water supply in the city as a whole, in parts of the town, or at the filling station is limited. On top of the chronic supply problems, moments of acute shortage or total absence of water supply can cause the water supply in the city to be cut off. As suggested by the list provided in Table 8.1, which covers only a few randomly selected periods over a span of 10 years, these shortages are not exceptional, but rather occur regularly, plunging the city into a condition of acute water crisis. The ever-present potential for moments of intense water shortage clearly ensures that the water issue remains high on the agenda.

Such times of serious shortage, even for those who are connected to the official water supply system, create conditions of speculative water fever that are invariably seized upon by the water vendors. It is quite common for water vendors to arbitrarily increase the price whenever a water shortage erupts. For example, when the water supply was halved on 28 April 1992 as a result of clogged filters because of abundant growth of water lilies (partly as a result of run-offs of nitrogen fertilizers into the River Daule), the tanqueros increased the water price by 50 sucres. These outbursts of more acute scarcity produce an environment that allows the tanqueros to appropriate an even greater monopoly rent from vending water. In 1998, during severe shortages, water vendors also arbitrarily increased water prices (up to US$60 for a truck load of 8 m^3). Given this power of water, it should not be a surprise that the water vendors actively use water scarcity as a strategic instrument to enhance their economic power. The production of water scarcity is, therefore, a preferred strategy for increasing the price of water (see Plates 8.2 and 8.3).

Table 8.1. Reported moments of acute water shortages in Guayaquil, January 1988–August 1993, May–August 1996, and January 1998–July 1998 (randomly selected periods)

Date	Place	Problem	Source	Stated Reason
11/01/88	Guayaquil	Low supply	*El Telegrafo*	Works at La Toma
03/07/88	Mapasingue	No water	*Meridiano*	Damage to mains pipe
12/08/88	Duran	No water	*Meridiano*	Part of mains pipe stolen
21/08/88	Guayaquil	Rationing	*Meridiano*	Maintenance works at electrical power station
12/09/88	Duran	No water	*Meridiano*	Part of mains stolen
04/10/88	South City	No water	*Telegrafo*	
15/01/89	Guayaquil	No water	*El Universo*	Works
13/02/89	Guayaquil	No water	*El Universo*	Water was diverted to the Peninsula for the tourists
03/03/89	Duran	No water		
08/04/89	Guayaquil	No water	*Expreso*	Lack of aluminosulphate
24/04/89	North City	Water shortage	*El Universo*	
20/06/89	La Toma	Low capacity	*El Telegrafo*	Pollution of river
21/07/89	Salinas	Water shortage	*La Segunda*	
03/11/89	Salinas	No water	*El Telegrafo*	Seven days of scarcity
15/11/89	Guayaquil	No water	*El Universo*	Sabotage and lack of chemicals to purify water
25/11/89	Guayaquil	Water shortage	*La Segunda*	Third week of scarcity
03/12/89	Guayaquil	Water problems	*El Universo*	La Toma did not have electricity
28/12/89	Guayaquil	Water scarcity	*El Universo*	
27/07/90	Guayaquil	Low pressure	*La Razon*	Growth of algae
21/09/90	Guayaquil	No water	*El Universo*	Electricity works in La Toma
25/02/91	Peninsula	No tanqueros	*El Universo*	Water delivery strike
27/05/91	Guayaquil	No water	*El Universo*	Installation of new pumps
19/08/91	Guayaquil	Water shortage	*El Universo*	Electricity problem
28/08/91	Guayaquil	No tanqueros	*El Universo*	Strike of 'tanqueros'
28/09/91	Guayaquil	Low pressure	*El Universo*	
07/11/91	Guayaquil	Shortage	*El Telegrafo*	Water works
14/11/91	Centenario	No water	*El Universo*	
18/11/91	Guayaquil	Shortage	*El Universo*	Works on mains
10/12/91	Guayaquil	Low supply	*El Universo*	
31/01/92	Centro/Sur	No water	*El universo*	Works
11/04/92	Guayaquil	No water	*El Universo*	Work
27/04/92	Guayaquil	Low capacity	*El Universo*	Problems with turbidity, lack of filters/polimers
20/05/92	Los Ceibos	No water	*El Universo*	Sabotage mains
20/05/92	South City	No water	*El Universo*	Sabotage mains
27/08/92	Guayaquil	No tanqueros	*Expreso*	Strike of 'tanqueros'
14/09/92	Guayaquil	No tanqueros	*El Universo*	Strike of 'tanqueros'
14/10/92	Los Ceibos	No water	*El Universo*	Works
Nov. 92	South City	Water shortage	*El Universo*	Electricity problems
18/01/93	South City	No water	*El Universo*	Electricity problems
13/03/93	Centro/Sur	Water shortage	*El Universo*	Damaged mains
24/03/93	Guayaquil	Water shortage	*El Telegrafo*	Increased turbidity
29/03/93	Guayaquil	Water shortage	*El Universo*	

Table 8.1. *Continued*

Date	Place	Problem	Source	Stated Reason
09/05/93	South City	No water	*El Universo*	Strike of 'tanqueros'
11/05/93	North City	Shortage	*El Telegrafo*	Problems with electricity
13/05/93	Guayaquil	No water	*El Telegrafo*	Damaged mains
31/05/93	Centro/Sur	No water	*El Telegrafo*	Damaged mains
01/06/93	Centro/Sur	Shortage	*El Telegrafo*	Electricity problems/ maintenance
07/06/93	Guayaquil	No water	*El Telegrafo*	Maintenance
11/07/93	South	No water	*El Telegrafo*	Connecting new mains
09/08/93	Guayaquil	Water shortage	*El Universo*	Damaged mains
16/05/96	Guayaquil	Cut-offs	*El Universo*	Damaged pipes
18/06/96	Guayaquil	No water	*El Universo*	
05/07/96	Guayaquil	Shortage	*El Universo*	Insufficient pressure
05/08/96	Norte	No water	*El Universo*	Repair works
02/10/96	Guayaquil	No water	*El Universo*	Energy shortage
16/01/98	Guayaquil	Shortage	*El Universo*	No chemicals
18/01/98	Guayaquil	No water	*El Universo*	Turbidity/'el Nino'
29/01/98	Guayaquil	No water	*El Universo*	Valves closed
01/02/98	Region	Shortage	*El Universo*	Broken mains
12/02/98–17/02/98	Guayauil	Shortage	*El Universo*	Broken pumps
06/03/98	Guayaquil	Shortage	*El Universo*	Damage to reservoir
10/03/98–20/03/98	Guayaquil	Shortage	*El Universo*	Combination of reasons
21/03/98	Duran	No water	*El Universo*	Turbidity
23/04/98	Guayaquil	Scarcity	*El Universo*	
07/05/98–17/05/98	Suburbio	Scarcity	*El Universo*	Maintenance
08/06/98–14/06/98	Guayaquil	Shortage	*El Universo*	Reduced production

Sources: Press reports from *Meridiano, El Universo, El Telegrafo, Comercio, Expreso, La Segunda, La Razón.*

8.3.2 Water strikes: manufacturing water scarcity

Indeed, the monopoly control over water distribution held by water vendors enables them to dry out large parts of the city and effectively dehydrate the citizens. In Guayaquil, this is the preferred strategy used to force through a significant price hike. Depriving the city of water for a few days allows the tanqueros to increase the price arbitrarily. Water price increases of more than 100% are not unusual immediately after a 'strike' by water speculators. Table 8.2 lists five episodes of intense confrontation and water strikes organized by the tanqueros in order to increase the water price over the period 1991 to 1993 (see also Plate 8.3). The ritual is invariably the same, with water vendors refusing to distribute water in the water-deprived areas, starting on a Friday morning, resulting in the people being deprived of water for three full days. These

Plate 8.2. Guayaquil's enduring water shortages and problems: media representations.

"Los tanqueros abusan de nuestras necesidades y nos cobran hasta 600 sucres", afirman dirigentes de barrios suburbanos.

El gran negocio de dueños de tanqueros

Crisis de agua sigue latente

Se han proliferado los sapos

Aso. de tanqueros denuncia la especulación con venta de agua

Concede EMAP-G

Accion Popular Contra Especuladores de Agua

En Guayaquil no hay agua, porque a mucha gente no le conviene que haya...

Tanqueros subirán precio de agua

Los vehículos que distribuyen agua potables a los suburbios elevarían el tanque de 55 galones a 1.300 sucres

Especulación de tanqueros ante escasez de agua potable

Escasez de agua provoca ola de especulación en Península

Suben precio al tanque de agua

Una elevación de 800 a 1.000 sucres actualmente imponen con arbitrariedad los tanqueros que distribuyen líquido vital a la isla Trinitaria, lo que ha originado la protesta de las familias humildes de este sector popular que necesariamente deben abastecerse de líquido vital.

Venden tanque de agua hasta en 2.200 sucres

Moradores del Guasmo Sur y de la Isla Trinitaria, denunciaron que el costo del tanque de agua subió de 1.200 sucres a 2.200 a partir del jueves pasado.

Venden tanque de agua hasta en 2 mil sucres

Innumerables abusos denuncian usuarios

Agua por tanqueros: otro vía crucis que padece el pueblo

Plate 8.3. Speculating with water.

Table 8.2. Examples of reported water strikes by the 'tanqueros', 1991–1993

Date	Source	Reason
25/02/91	*El Universo*	Strike of 137 'tanqueros' on the Peninsula after the police arrested 17 'tanqueros'
28/08/91	*El Universo*	Strike of 'tanqueros' after the police had arrested two who sold water without a permit
27/08/92	*Expreso*	48-hour strike of 'tanqueros' because EPAP suspended and fined fifty of them
14/09/92	*El Universo*	'Tanqueros' refuse to distribute water for three days after government announced price increases for gasoline
May 1993	*El Universo*	'Tanqueros' go on strike after the relocation of the filling station from Bellavista to Via a la Costa 10 km

Sources: Interviews, newspaper reports.

actions also usually take place in the midst of the dry season in which no alternative (pluvial) water source is available. Water is literally wrung out from parts of the city in an inevitably successful attempt to jack up the price and increase the money rents that can be extracted.

Moreover, water strikes are also used as a means of defending the collective interests of the water vendors and as an expression of solidarity. This is a particularly effective weapon against state attempts to regulate, intervene, or punish water vendors who charge too high a price or distribute water without the necessary permits. Table 8.2 also lists a number of occasions when the weapon of water strikes was used to counter actions undertaken by the police against some of the water vendors. The danger of civil unrest resulting from halting water distribution and the potential tensions arising from such situations invariably leads to the state backing down on its demand (albeit sometimes tacitly), and accepting the new demands of the water distributors.

Changes in the location of the official filling stations for the tanqueros is also used as an excuse for increasing the price. Between 1990 and 1993, the filling stations changed location three times. In April 1990, EPAP decided to concentrate the filling stations to a limited number of places. Before that time, there were more than 20 filling points, mostly located along the Avenida 25 de Julio, which runs from the city centre to the new port. This decentralized system prevented the authorities from controlling and accounting for water sales to the water distributors, and led to widespread theft. Two new filling stations were constructed, one at Bellavista (to supply Guasmo and Suburbio) and one at Via a Daule to supply the Northern settlements. While this concentration allowed better control and supervision of the tanqueros, it also resulted in a price rise as the water vendors claimed that the transportation distance had increased and, consequently, delivery costs went up. At the time, the price rose from 150 sucres to 200 sucres, a 33% increase. In late August 1991, EPAP-G refused to let the tanqueros use Puente 17 which allows the lorries to take the shorter road to Guasmo, and seven tanqueros were penalized for failing to obey the regulation. In response, the other tanqueros went on strike on a Thursday, drying out the city, and the Frente de Usuarios organized a march to protest against the water scarcity. EPAP-G allowed residents to fill up with water at the distribution station without charge in order to appease the masses, whilst at the same time changing the water tariffs charged to water vendors and increasing the price of water for domestic use from 15 to 70 sucres for 1,000 litres, thereby ending the price difference between domestic and industrial water usage. After the strike, the real stakes became clear. Retail water prices increased from 300 sucres to an average of 400–450 sucres. This suggests that the measures taken by the urban authorities immediately prompted a concerted response and action from the tanqueros, who then used their socio-economic power to pass on the cost associated with the changes to the final customers. It is these consumers who find themselves in a very powerless situation, with no other options than accepting the new conditions or dying of thirst.

September 1992 and May 1993 saw the tanqueros engaged in two dramatic moments of intense action. These indicate the considerable social power wielded by those who have a quasi-monopolistic control over the spatial distribution of water in the city. The deterioration of water supply during the early nineties, combined with the austerity and structural adjustment policies pursued by the newly elected government headed by President Sixto, caused overall living conditions to deteriorate rapidly. In order to stem inflation, suppress domestic demand, and promote a reinvigorated export-led development, the Sixto government announced and implemented a major economic restructuring plan in August/September 1992. Wages were frozen, while prices for basic food and other commodities and energy rose spectacularly (between 50% and 160% (for gasoline)). The tanqueros went on strike on a Friday to protest against the price increase of gasoline and spare parts, but this interruption of water deliveries was also, as became clear a few days later, part of a strategy to increase the price of water to the consumer. The ensuing popular unrest among those who no longer had access to water took the form of rebellions, attacks, and rioting. Tanqueros became subject to attacks and the army had to be called in to restore order, with water distribution finally resuming on the following Monday morning. This strike led to a price rise from 450 sucres to 700/800 sucres (see *El Universo* 14–17 Sept. 1992).

Less than a year later, in May 1993, the tanqueros were once again at the forefront of an intense struggle. The local government had once again decided to relocate the filling station of Bellavista from its central location to a site near the beltway on Via a la Costa km 12.5. The official reason given was to improve the accessibility of the site for lorries supplying Guasmo and Suburbio. In reality, however, the municipality was launching a major clean-up campaign in anticipation of the upcoming Copa Americano (South American international soccer tournament), a major international event which was to be co-hosted by Guayaquil. The filling station of Bellavista (which supplies Guasmo and Suburbio) was located very near to the main soccer station where some of the key games were to be played. The event was seen to have important public relations implications, and cleaning-up the area was considered essential to boost the image of Guayaquil. In 1990, the Bellavista filling station cost 151 million sucres to build, but was closed after just three years of operation. The new station opened on 4 May 1993, but, at that time, only 16 of the planned 26 supply lines were operational, resulting in long waiting times (of about 1½ hours) and waiting tanker lorries queuing for several kilometres. On 5 May, EPAP-G declared that the water price would not rise because the distance between the filling station and the serviced areas is shorter than before, whilst the tanqueros 'investigate' the possibility of a price rise. On 7 May 1993, the tanqueros went on strike to protest against the long waiting times, which reduced the daily number of trips they were able to make from 7 or 8 to not more than 3 or 4. They estimated that 2,000 trips were needed daily to supply Guasmo and Suburbio, while the new filling station only allowed a daily maximum number of about 900 rounds, and announced a price increase for a 200-litre barrel of water from 800 to 1,200 sucres. This strike continued until at least 11 May, resulting in water delivery to Guasmo and Suburbio being reduced to a trickle or completely stopped for almost an entire week. The few tanqueros that did make water-vending trips demanded prices of up to 1,500 and 1,800 sucres (almost US$1) for 200 litres. Most of them drove in from the filling station at the Via a Daule where business was going on as usual. People who were becoming increasingly desperate to get water convened at the filling station to demand it, hired private trucks, and formed informally organized neighbourhood hauling parties to get at least some water.

The conflict continued throughout the entire month of May. On 11 May, in a desperate attempt to defuse tension, EPAP announced that it would close down industrial water supply lines in the southern part of the city to defuse the rapidly intensifying social tension. By early June, the situation had still not returned to 'normal' as water remained extremely scarce in the central and southern part of the city. On 31 May, a march was organized to protest against the situation, and tanqueros had to be brought in under police supervision and directed to areas with extreme shortages. Extraordinary high water prices were charged, with 200 litres now costing up to 2,000 sucres (1US$). The social struggle intensified, with groups of people taking to the streets, incinerating tyres, blocking roads, and occupying the main road connecting the central part of the city with the port. The Director-General of EPAP-G, Mario Chavez-Baird, personally supervised the closing down of illegal connections to show that official action was indeed being taken to redress the situation. On 7 June, the water shortage was still acute, with marches and actions continuing (see *El Universo*, *El Comercio*, and *El Telegrafo* 4–11 May/2–4 June). By mid-July, FEDEBAS demanded the establishment of a local management committee that would involve all the key actors in an attempt to negotiate a more regulated supply of water. By now, the strike had ended but water supply remained limited and problematic. The 'new' water price settled at around 1,200 sucres for 200 litres (a 50% increase), with a higher rate of 1,500 sucres charged in the most distant areas (such as Isla Trinitaria).

8.3.3 Striking the water system: 'Una Guerra por el Control de los Recursus Hidricos'

In their ongoing rivalry with the water utilities, the water speculators do not shy away from terrorist attacks on the official network to prevent the erosion of their monopoly position or to expand the geographic area over which they hold the key to life and death. When the distribution network in Guayacanes, Los Ceibos, and Sauces was sabotaged in May 1992, *El Universo* reported the event under the heading of 'Una Guerra por el Control de los Recursus Hidricos' (a war for the control of water resources). The enormous money power associated with the control over nature can therefore turn into an outright war over water. A traditional point of attack is the mains pipe connecting the city of Guayaquil with the coastal resort towns of the Peninsula, in particular Salinas. These acts of sabotage invariably take place during the 'winter' months (the rainy season), during which the environmental and climatological conditions in the city are such that thousands of the better-off residents escape from the hot and humid conditions in the city and retreat to their coastal residences. Salinas is nevertheless dependent on the same water supply system as Guayaquil and is connected with the main reservoir by means of just one aqueduct. Sabotaging the aqueduct ensures an instantaneous expansion of the market for the private water vendors and guarantees a spatial monopoly over water distribution on the Peninsula. When the aqueduct was sabotaged in November 1989, the price for a truckload of water rose from 10,000 to 14,000 sucres. Table 8.3 lists a number of reported sabotaging events on the water system (see also Plate 8.4).

Table 8.3. Examples of reported attacks and sabotages of the water system, 1988–1992, February–June 1998

Date	Source	Nature of attack	Place
12/02/88	*Meridiano*	Breaking and stealing pipe	Duran
03/11/89	*Telegrafo*	Attack on mains pipe	Salinas
Feb. 91	EPAP-G	Sabotage aqueduct	Guayaquil/Salinas
13/03/91	*Telegrafo*	Attack on mains pipe	Between Duran and La Puntilla
20/05/92	*El Universo*	Sabotage distribution pipes	Guayacanes, Los Ceibos, and Sauces
08/02/98	*El Universo*	Broken mains (reason unknown)	Peninsula
08/06/98	El Universo	Accusations of sabotage made by community organizations	

Sources: Newspaper reports, EPAP-G documentation.

Desabastecimiento de agua por paro de los tanqueros

Los sectores suburbanos y marginales del sur de la ciudad soportaron el desabastecimiento de agua potable, por la paralización de los 220 tanqueros que distribuyen el servicio desde la bocatoma del kilómetro 10 1/2 de la vía a la costa. Estos anunciaron que el costo por tanque será de 1.200 sucres.

Tanqueros "terroristas"

Tanqueros paralizan labor

Pocos tanqueros distribuyen agua

Responsables serán drásticamente sancionados

Sabotaje causó daños en tubería de agua potable

Autoridades investigan

Sabotean acueducto

Sector norte sin agua por manipulación de válvulas

Plate 8.4. Going on water strike/sabotaging the water system.

8.4 Conclusions

In short, those in control of water must be perpetually involved in a generalized and overt water class struggle in order to maintain the basis of their power. The tactics, strategies and struggles waged by those that hold effective political and/or economic control over water production and distribution point towards an outright social conflict over the urbanization of nature. We have shown how the productionist logic of the water authorities provided the elites with exclusive access to unlimited amounts of water while ignoring the key issue of distribution. The strategies and policies of international lending agencies also permit the 'urban water managers' to safeguard their exclusive control, to sustain a productionist logic, and to allow the urban elites to withdraw as much water as they desire at low prices. The clientelist political system and paternalistic political relations, in turn, prevent those deprived of water from turning these issues into coherent class politics by providing individualized and particularized solutions, which are granted in a piecemeal and highly politicized fashion. This system is self-perpetuating, reinforcing the dependence of the water-deprived citizens on clientelist ties and patronage.

Although the state's actions are influenced by threats of urban revolt and discontent, which threaten the political economic cohesion of the urban fabric, it continues to allow the perpetuation and consolidation of the private truck-based distribution system. These strategies become more and more direct and oppressive, turning the water sellers into central power brokers in the city. In turn, this allows them to monopolize water distribution and to cash in on their power position by means of appropriating extraordinarily high water rents. This water war meets generally with a fragmented, rather weak, and often individualized popular resistance.

All of this confirms that water and water issues are central to understanding the political ecology of the city. It also shows how the contestation of nature's water is directly related to the sustainability of urban life. In the end, it begs the question as to how an emancipatory and empowering water politics can be achieved.

PART III

Conclusion

9

Whose Water and Whose City? Towards an Emancipatory Water Politics

9.1 Setting the agenda: nature, justice, and the city

Urban water is part and parcel of the political ecology of power that structures the functioning of the city. The preceding chapters showed how the circulation of water, the most natural of goods, is inserted into the maelstrom of social power relations through which the urbanization process unfolds. The urbanization of water and the urban hydrosocial cycle on which the sustainability of the city depends is impregnated by myriad social, cultural, political, and economic meanings and powers that place control over and access to nature squarely within the realm of key moral and ethical questions. In particular, it raises the issue of the relationship between nature, social justice, and the city. While the city cannot exist without the perpetual metabolic transformation of nature, this very transformation turns nature into a deeply social process in which nature, society, and the city can no longer be separated. It also suggests, of course, that some key questions need to be asked with respect to the social and material construction of the city's nature and the power relationships through which this transformation and urban metabolism is organized and maintained. In this vein, we have attempted to explain how a political ecological analysis permits casting a new and different light onto the socio-ecological metabolic dynamics of the urbanization process itself.

The previous chapters indicated that this urban transformation of water is a manifestation and expression of wider relations that clearly transcend the simple question as to who does and who does not have access to water. It also suggested that the water problem is not merely a question of management and technology, but rather, and perhaps in the first instance, a question of social power. The many manifestations of this power discussed in this book suggest how an enabling and empowering water politics need to address this question of power head on. In particular, those that defend the rights of the disempowered to the city's nature have to understand the central political power relationships that structure the existing pattern. It is exactly this exclusion that needs to be contested if a genuinely emancipatory and fully humane urban development is to be constructed.

Clearly, the analysis presented in this book aims to reposition the issue of water availability into the broader framework of urban politics. Every feasible or even mildly effective solution to the thorny problem of water control and access is doomed to fail unless it recognizes what it is up against and what is at stake in this struggle. In the end, this whole problematic raises the issue as to who has the right to the city and whose nature is, in the name of progress and modernization, so violently and oppressively appropriated by some at the exclusionary expense of others.

9.2 Towards an emancipatory urban water politics: principles

Although the urbanization of water is systematically couched in the rhetoric of engineers and technicians, it should be clear by now that it is difficult to sustain the argument that this is a predominantly technical matter. Technology clearly matters, but the technology itself is embedded in and is an expression of a wider political discourse and practice, which then becomes *de facto* built into the steel and concrete of the technological engineering structure (Kaïka and Swyngedouw 2000). However, the technocratic discourse and argumentation help to obfuscate the issues of power and control and to formulate the water problem as one that is determined by the power over nature rather than by the power of one social group over another. We have shown how the power of one social group over another is mediated by and organized through the ways in which nature is transformed and socialized.

It is quite clear that the technological argument belongs to the arsenal of discourses, if not ideologies, developed and advocated by those who hold power. Not surprisingly, the technological system itself helps both to maintain the mechanisms of control and exclusion while simultaneously contributing to the construction of an argument that ignores or, at best, minimizes the role of socio-political power. It is therefore scarcely surprising to discover that those who control the transformation of water, whether they be the state and its representative organizations or private water supply companies, view this as an exclusively technological exercise.

Elements for an emancipatory and empowering water circulation system include managerial, production, distribution, and supply considerations. In the first instance, it is imperative that the notion of unlimited supply of water be replaced by a view of water management as the just distribution of an inevitably limited, but vital, resource. However, the urban 'scarcity' of water is not a function of its limited availability in nature, but rather, as has been shown, is socio-economically and politically constructed. And it is, of course, exactly this manufactured 'scarcity' that permits water to enter easily into the unequal power relations associated with market relations, and to turn the control over H_2O into a means to exercise bio-political power and control (Hardt and Negri

2000; Crespo 2002*a*). A managerial structure concerned with distributional, rather than predominantly production related, issues is, therefore, of vital importance. This would question the existing power relationships in the city, as well as raising the critical issues of 'Whose water' and 'Whose city' we are talking about. Challenging the systematically uneven water power that characterizes the city of Guayaquil, or any other city for that matter, requires closer involvement of the local people in the management of and control over the public water utility. This improved participation of the citizen and the reversal of the productivist logic could lead to a reorientation of water politics based on considerations of equal distribution and just accessibility. The pricing system ought to be adjusted to more accurately reflect the purchasing power of the water consumer. This would necessitate bringing water distribution in the marginal settlements under direct control of public authorities or of collective grass-roots organizations. In fact, the appropriation of water rents, currently held and monopolized by the private water vendors, would easily allow for an effective and future oriented urban water politics and economics.

Clearly, the transformation to an emancipatory urban water politics is not only a local or urban issue, but is also inserted into the higher political scale of the state and the international political economy. All of this has profound implications for international and national financing schemes, the managerial structure of the water utility, the financing of water schemes, and the politics of privatization. Moreover, fundamental changes would be required in the technological organization of the water production, conduction, and distribution system itself.

Those that are disempowered and excluded, however, have long understood the political nature of the water problem. Women, indigenous populations, and the poor experience the deeply exclusionary practices of unequal water accessibility in their daily quest for water, and realize that this is structured through social power and not through the technologies of water mastering. It is in this sense as well that we wish to develop a series of suggestions. We do not intend to propose an alternative Master Plan for urban water supply. Rather, we wish to emphasize how the water issue needs to be dealt with as a political problem in the first instance. The technological discourse and practice, and the socio-physical metabolism of the city, need to be grafted onto an emancipatory urban water politics.

9.3 Towards a new water politics: from the local to the global and back again

9.3.1 A voice for the grass roots

Among the poor and excluded, the water issue is a top priority and clearly one that is formulated in outright political terms. Although apparently the least powerful, on occasion they can muster sufficient force to resist further attacks on the supply system or to demand institutional, organizational, or infrastructural changes. Grass-roots organization is a cornerstone for an emancipatory urban living and needs to be structured around some key premises.

Sustained grass-roots activism

As many neighbourhood organizations have demonstrated in Guayaquil and elsewhere, sustained and organized popular action has helped to keep the water issue high on the political agenda and forced the official system to remain on guard. A variety of strategies have been employed, ranging from marches and rebellions to the formulation of alternative water distribution and production strategies (Crespo 2002*a*, *b*). When combined with a sustained criticism of the official attitude towards the water speculators, these strategies have contributed to turning the water issue into a clear and key political theme that can no longer be addressed primarily in terms of engineering principles, investment schemes, and technocratic solutions. FEDEBAS, together with a host of other organizations, have managed to keep the attentions of local, provincial, and national government on the water issue, with a fluctuating level of success. Although direct tangible results are difficult to measure, their relentless defence of the excluded and urban underclass has been pivotal in the city's political economy.

In addition, women's organizations have long recognized the power relationships associated with and embedded in the metabolism of urban water. Their strategies, preferably in close association with other grass-roots movements, are of prime importance in keeping the issue in the spotlight. Such sustained grass-roots action which gives a voice and a face to the underprivileged, is a necessary condition for moving in the direction of a more sustainable urban political ecology.

Negotiating power

In addition to raising their concerns and demonstrating their potential for action, it is imperative for people's organizations to become directly involved in the planning, implementation, and management of the water production and distribution system itself. This means pressing the water authority to place the issue of water distribution (rather than production) at the top of the agenda. This would necessitate the integration of the truck-based distribution system into the policy framework. While the current policy system still officially holds to the view that truck-based distribution is a temporary anomaly that will disappear as the official piped network expands (an idle promise recently reinvigorated by the privatization debate), the neighbourhood organizations have to press the water authority to accept the inevitability of a truck-based delivery system for a long time to come. Recognizing alternative distribution systems as inevitable would bring the regulation and organization of the private vending system under the remit of the state and/or the water-regulators, who would then have to either plan and implement alternative distribution channels themselves or insist on a more strict regulation of the existing private water distribution system. This system would include the establishment of democratic and participatory bargaining institutions or frameworks in which all those concerned could negotiate key water-related issues (price, quality, distribution planning, etc.) which would, in turn, be appropriately enforced.

Implementing grass-roots projects

In addition, such a system of bargaining power needs to be extended to incorporate planning and implementation issues such as the financing and construction of community taps and 'cisternas comunitarias', the planning of new caption sites and alternative supply systems, or the planning and implementation of alternative smaller-scale production systems (using, for example, subterranean water as some of the hotels and local industries do). Finally, neighbourhood organizations can and should take the lead in organizing alternative distribution systems. By reinvesting the profits obtainable from these in community development projects, the financial benefits could be retained locally rather than siphoned off to Miami bank accounts, the vaults of the global water industry, or the share-holders of water companies.

For example, a feasibility-study undertaken in September 1993 by FEDEBAS and myself suggested that an initial investment of 700 million sucres to buy five tank lorries for the distribution of water by the community in some settlements of Suburbio would generate a net annual profit of 34.7 million sucres, while the price of water could be kept down to 700 sucres for a tank of 200 litres. This would mean a 40% price reduction to the consumer, allow a massive annual reinvestment into the community, and facilitate the full-time employment of 13 community-based people (Swyngedouw 1994; Vlaams Internationaal Centrum 1994). This suggests that a minimum investment can yield high returns while addressing the thorny issue of water scarcity and transfer of financial resources as a result of the speculative strategies of the private water vendors. This project is currently funded by an international NGO in cooperation with FEDEBAS. It was launched in 1995 and the first two water trucks, financed by the Belgian state, arrived in Guayaquil in March 1996. Another three trucks arrived a few months later. The implementation phase of the water distribution scheme was put in place and the full project was operational by the end of 1996.

9.3.2 The local state

The role of the local state (urban and provincial) is central to the success of such a transformation. As privileged interlocutors between communities and water company, the local state has the power to place pressure on the water authorities and company, to negotiate power with citizens and water owners, and to project, promote, implement, and police a vision of urban development and management in which emancipation and empowerment are central concerns.

In addition, the inequalities of the current and highly skewed distribution system need to be tackled. This will require addressing technical design issues (such as bypass systems enabling a more spatially just rationing of the available water), as well as implementing a pricing system that seriously sanctions spillage and massive overconsumption of water for conspicuous consumption purposes (fountains in gated communities, swimming pools, irrigation of gardens, etc.). Perhaps the most important issue, however, is the recapture of the enormous economic water power currently held by the tanqueros. The willingness-to-pay for water in the suburban areas is understandably very high, and (as shown in Chapter 7) the total revenue of the private water vending industry is higher than that of the water utility. Taking control of distribution in these settlements would cause an immense improvement in the economic and financial position of the water company, creating a surplus that could then be ploughed back into a further improvement of the distribution system. In addition, the highly subsidized pricing of water for the well-to-do parts of the city reinforces the injustices, creating a system of cross-subsidization that favours the rich and has clear socially perverse effects. Redesigning the pricing system in a way that would place the ability-to-pay in the foreground would equally be a key element in altering the socio-spatial pattern of water distribution.

Finally, the productivist logic combined with the ideology of underdevelopment that so successfully contains the local tensions and contradictions by exporting them needs to be replaced by a determined sensitivity to local power relationships and how these define water accessibility and exclusion. This all hinges on the capacity to regulate and police access to nature, although in the current power situation control is virtually absent or plainly impossible. The current intricate power rituals around water control and accessibility, combined with the exclusion of many segments of the population in the decision-making process, preclude the successful policing of and control over water accessibility. Populist political strategies are always present, and prevent the adoption of measures that empower large segments of the population and aim at a more fundamental reorientation of public service provision.

9.3.3 The national state

Nature and the environment are not 'external' conditions, independent of the urban process, but are part and parcel of the political basis of social change. The economic, social, ecological, and infrastructural processes inherent in urbanization require the simultaneous transformation of nature and the construction of a new socio-natural environment. The state's technocratic and economistic perspectives ignore these crucial connections, thereby perpetuating profound injustices and undermining the sustainability of the social and urban system. This is reinforced still further by the demands of the international funding and lending agencies. Environmental justice, as explored through the urbanization of nature's water, is essential to the long-term survival of the national economy and the sustainability of its social and environmental fabric. The killing of thousands of citizens as a result of problematic water access loots the country of some of its most vitally needed resources. It has to be recognized that this loss of life and the disempowerment of those living on the urban ecological and social margins is not a poisonous 'gift of nature' but the result of concrete political ecological transformations structured through the political, cultural, and economic power positions of the elites. Turning political economy into political ecology with a strategy inspired by considerations of socio-environmental justice is essential if a positive and enabling interweaving of society, nature, and the city is to be regained. This in turn requires a systematic resistance to the dominant forces that celebrate a continuing ecological colonization, international control over and private appropriation of nature's resources, and the perpetuation of bio-political command structures.

Little is to be expected from the current neo-liberal export-based, and structural adjustment policies pursued by the national government. If anything, the shift in expenditure patterns and policy initiatives over the past few years have signalled a systematic shift away from a commitment to the poor and powerless, and towards supporting a reinvigorated internationalized local and global elite. Much of this struggle unfolds over resources (such as oil, water, and land) and their use. This is apparent in the many highly contested cases of water privatization around the world, including that of Guayaquil's urban water supply and sewerage company. The case of Cochabamba, Bolivia, indicates that the contestation of the regulation of nature can be successful if mass mobilization threatens the cohesion of civil society.

In short, national policy has thus far focused almost exclusively on mechanisms to intensify the transformation of nature (and, consequently, the production of a new nature) without much consideration of the issue of distribution. The further mastering and control of nature, however desirable it may seem from a technocratic engineering perspective, does not necessarily contribute to a more sustainable and socially just form of urbanization and urban development. Clearly, taking the distributional issue into account (that is, who has the right of access to transformed nature) is not independent of a consideration of the ways in which nature will be transformed and socialized. Such a reorientation will, in turn, affect the position of Ecuador in the international division of labour and the relations between the state and international financial and other organizations.

9.3.4 Mobilizing water and the politics of scale

The urbanization of water is, of course, caught into the wider political economy of the state and the international position of Ecuador. The recent shift towards a more liberal monetarist policy, which has resulted in a significant reduction of expenditures in social domains (see Acosta and Maldonada 1992; Suarez-Torres *et al.* 1997), has had negative effects on urban water availability and sanitation. In addition, the structural adjustment policies advocated and often imposed by the World Bank and the IMF have not only reduced the financial flows from the capitalist core, but also has made them subject to stringent conditions which slow down the expansion of infrastructure and put a premium on privatization, efficiency, productivity, export-led growth and, above all, profitability. This, in turn, further erodes the possibility of supplying the poor. In addition, the reinvigorated export dependency pursued to redress the balance of payment problem also affects social and physical infrastructure investments for the poor by emphasizing investments in and promoting ventures aimed at increasing foreign revenue earnings. The ecological transformations associated with these internationally policed bio-political restructuring policies insert the political ecology of Ecuador even more squarely into the international socio-ecological division of labour at the expense of the transformation of nature for local needs. This is not a new phenomenon. Chapters 4 and 5 showed how this has been the fate of Ecuador since the nineteenth century and is integral to the way in which Ecuador became inserted into the international division of labour during the twentieth century. In other words, water and other parts of nature are turned into monetized and globalized nature rather than into a milieu in which a productive and mutually supportive relationship develops between a just urban sustainability and a responsible management of nature's resources.

Grass-roots movements should recognize these linkages and scale issues by putting them onto their own agenda. Without a systematic debate over the wider political ecology of the urbanization process, the people's projects will remain at the margins of a transformation process whose contours are drawn by national and supranational institutions and power brokers. In the international domain, water is indeed increasingly becoming a problematic resource (see Postel 1992) that is fought over in innumerable ways by communities, states, and international water companies alike. Clean and potable water (as well as irrigation water) is rapidly becoming the new 'Blue Gold' over which an intense geopolitical struggle unfolds. This of course raises the issue as to who has the right of how much water of what quality and where. For international organizations and financial institutions, the transformation of nature is still largely considered in terms of the balance sheet of cost–benefit analysis in which the short-term monetization of nature and the discounting of future costs are the guiding principles. Such assessment mechanisms further reinforce the exclusion of the poor. In particular, the deafening calls for privatization of urban and public services sounds rather hypocritical in an environment like Guayaquil in which 40% of the population is actually serviced by private service providers who earn more than half the total water revenue. The structure of private water provision in Guayaquil demonstrates the potentially lethal effects of the privatization of nature. It also suggests that private organization is not necessarily a panacea for remedying unsatisfactory service provision.

In addition, international financial flows have been jeopardized by Ecuador's problematic history of debt repayments, and have turned from an abundant flow during the oil-rich years of the 1970s to a mere trickle in the late eighties and early nineties. Moreover, those thin flows that do remain have become increasingly subject to stringent conditions, which usually come down at the expense of the poorest segments of society. These conditional loans also change the society/environment interaction, both in terms of the type of environmental transformation that is pursued and the means used to facilitate this. These policies are unlikely to change in the foreseeable future unless the international community increases pressure on bilateral and multilateral financial assistance to incorporate considerations of socio-environmental sustainability and justice.

9.4 Technology, infrastructure, and the appetite for water

In the end, it is the technological and engineering practice and systems that legitimize a continuation of the current dominant vision and strategy. Of course, this strategy then becomes further cemented into a technological organization of the process of transforming nature. For example, tackling the distributional issue would also necessitate a reversal of the technological/engineering structure that characterizes the current system, which is aimed at maximizing water production. The enormous spillage, skewed pricing structure, and highly exclusive distribution system suggest that the other side of the circulation of water is under-studied, undervalued, and under-resourced.

It is fairly simple to devise alternative technological systems that would embody a different set of social relations and have positive social implications. The southern suburbs of the city, for example, could easily be given a better supply of water by constructing a mains pipe directly from the reservoir at Tres Cerritos to the poor suburban settlement of Guasmo. This would improve water pressure and supply in this area (but of course at the expense of the upper-income areas of Urdesa and Los Ceibos). In addition, tapping into alternative water reserves by means of small-scale technology (for example, small pumps using aquifer water) would increase the power of local communities over their own water supply. In particular, this would diffuse the power of the 'ideology of underdevelopment' argument. Although partially correct, the ideology of underdevelopment is used as a powerful tool to legitimate and explain away persistent water exclusion while maintaining exclusive control over water by the middle and upper classes. In addition, reversing the technocratic argument would weaken the role of international finance in the race to commodify and urbanize nature's water.

Addressing these key issues is essential in order to gain control over the wider and longer-term objective of reducing water losses and expanding water production capacity.

9.5. Whose nature? whose city?

In the end, the urban water issue is part and parcel of a much wider consideration of the environmental basis of the city's existence and change over time. The city is a giant social process, perpetually transforming the socio-physical metabolism of nature. Nature and society are in this way combined to form an urban political ecology, a hybrid, an urban cyborg that combines the power of nature with the power of class, gender, and ethnicity. The water issue illustrates how nature and society combine in the production of socio-spatial fabric that privileges some and excludes many.

Water, therefore, is an integral element of the political ecology of the city and needs to be addressed in these terms. Urbanizing water, although generally portrayed as a technological engineering problem is, in fact, as much part of the politics of life as any other social process. The recognition of this political meaning of nature is essential if sustainability is to be combined with a just and empowering urban development; an urban development that returns the city and the city's environment to the citizens.

BIBLIOGRAPHY

Acevedo, H. M. (1938). 'Estudio del Approvechamiento de las Aguas del Rio Daule para Abastecerde Agua Potable a la Ciudad de Guayaquil', thesis submitted for the degree of Civil Engineering, Faculty of Mathematics and Physics, University of Guayaquil.

Acosta, A. (1990). *La Deuda Eterna: Una Historia de la Deuda Ecuadoriana, Grupo de Trabajo sobre Deuda Externa y Desarollo*. Quito: ALOP/CAAP/CECCA/CERG/CIUDAD.

Acosta, J. (2002). 'Ecuador: Workers Rise Up Against Privatization in Guayaquil'. Quito: Centro de Derechos Economicos y Sociales (www.citizen.org/documents/ACFC0.pdf)

——and Maldonado, G. (1992). *Pobreza y Politica Social en el Contexto del Ajuste. Analysis Macroeconomico: El Caso del Ecuador*, Consultancy report (preliminary version). Quito: Instituto Latino-Americano de Investigaciones Sociales (ILDIS).

Aguirre, R. (1984). *Estado y Vivienda en Guayaquil*, Colección tesis, no. 4. Quito: Facultad Latinoamericano de Ciencias Sociales (FLACSO).

Allou, S. (1987). 'Introduccion Historica. Formas Urbanas y Formaciones Sociales en el Ecuador: Los Principales Actores', in *Geografia Basica del Ecuador, iii. El Espacio Urbano en el Ecuador*. Quito: Centro Ecuatoriano de Investigacion Geografica, 16–37.

Amin, A. (1994) (ed.). *Post-Fordism: A Reader*. Oxford: Blackwell.

Anton, D. (1993). *Thirsty Cities: Urban Environments and Water Supply in Latin America*. Ottawa: International Development Research Centre.

Arellano, M. E. (1992). 'La Verdad Sobre el Agua Potable'. *El Universo*, 12 April 1992.

Armstrong, W. and McGee, T. (1985). *Theatres of Accumulation*. London: Methuen.

Asociación de Tanqueros '25 de Octubre', Pre-Cooperativa de Transporte de Agua y Servicios de Basicos 'Independencia y Libertad'. 'Proyecto de los Costos Operacionales por Vehiculo', Guayaquil, 24 Sept. 1990 (mimeographed).

Ayala, E. (1982). 'Cacao, Capitalismo y Revolución Liberal', *Cultura, Revista del Banco Central del Ecuador*, 13: 91–125.

Bachelard, G. (1942). *L'Eau et les Rêves*. Paris: Editions Corti.

Baeten, G. (2000). 'Tragedy of the Highway. Empowerment, Disempowerment and the Politics of Sustainability Discourses and Practices', *European Planning Studies*, 8(1): 69–86.

Báez, R. (1985). 'Apogeo y Decadencia del Modelo Agroexportador: Periodo de la Segunda Posguerra', in R. Agoglia (ed.), *Historiografia Ecuatoriana*. Quito: Banco Central del Ecuador/Corporación Editora Nacional, 549–62.

——(1992). 'Visión de la Econom¡a Ecuatoriana, 1948–1970', in *Ecuador de la Postguerra: Estudios en Homenaje a Guillermo Perez Chiriboga*. Quito: Banco Central del Ecuador, 43–56.

Bakker, K. (1999). 'Privatizing Water: The Political Ecology of Water in England and Wales'. unpub. D.Phil. thesis, School of Geography and the Environment, Oxford University.

Bakker, K. (2000). 'Privatizing Water: Producing Scarcity: The Yorkshire Drought of 1995', *Economic Geography,* 76(1): 4–27.

——(2001). 'Paying for Water: Water Charging and Equity in England and Wales', *Transactions of the Institute of British Geographers*, 26: 143–64.

——(2002). 'From State to Market?: Water *Mercantilización* in Spain', *Environment and Planning A*, 34: 767–90.

——(2003). *An Uncooperative commodity: Privatizing Water in England and Wales*. Oxford: Oxford University Press.

Balibar, E. (1995). *The Philosophy of Marx*. London: Verso.

Bataillon, C. and Panabière, L. (1988). *La Plus Grande Ville du Monde: Mexico Aujourd'hui*. Paris: Publisud.

Batisse, M. and Grenon, M. (1989). *Futures of the Mediterranean Basin: The Blue Plan*. Oxford: Oxford University Press.

Beck, U. (1992). *Risk Society*. London: Sage Publications.

——(1995). *Ecological Politics in an Age of Risk*. London: Polity Press.

Benaiges A. M. (1991). 'Organización de la Población y el Servicio de Agua y Saneamiento', in Federación Mundial de Ciudades Unidas (ed.), Empresas Prestarias de Servicios de Agua y Alcantarillado y Usuarios. Quito: CIUDAD, 99–106.

Bennett, V. (1988). 'Women and Class Struggle in the Sphere of Reproduction: Social Conflict over Urban Water Services in Monterrey, Mexico, 1973–1985', Paper presented at the 46th International Congress of Americanists, 4–8 July, Amsterdam.

——(1995). *The Politics of Water: Urban Protest, Gender, and Power in Monterrey, Mexico*. Pittsburgh: University of Pittsburgh Press.

Benton, T. (1989). 'Marxism and Natural Limits: An Ecological Critique and Reconstruction', *New Left Review*, 178: 51–86.

——(1996) (ed.). *The Greening of Marxism*. New York: Guilford Press.

Bernal, P. I. (1991). 'Suministro de Agua en los Barrios Marginados de Barranquilla: El caso de los Carrotanques Particulares', in Federación Mundial de Ciudades Unidas (ed.), *Empresas Prestarias de Servicios de Agua y Alcantarillado y Usarios: ¿Pueden Tener y Desarrollar Proyectos Comunes? ¿Cuales? ¿Como?*. Quito: CIUDAD, 153–6.

Biswas, A. (1979) (ed.). *Water Management and Environment in Latin America*. Oxford: Pergamon Press.

——and Kindler, K. (1989). *Sustainable Water Development and Management*. Nairobi: United Nations Environment Programme.

Blowers, A. (1993) (ed.). *Planning for a Sustainable Environment*. London: Earthscan.

Bocco, A. (1987). *Auge Petrolero, Modernización y Subdesarollo: El Ecuador de los Anos Setenta*. Quito: Corporación Editora Nacional.

Bock, S. (1988). *Quito, Guayaquil: Identificación Arquitectoral y Evolución Socio-Económica en el Ecuador (1850–1987)*. Guayaquil: Corporación de Estudios Regionales, Guayaquil.

Bond, P. (1997). 'Privatization, Protest and Participation: Citizen Opposition to the World Bank in Haiti and South Africa', Paper presented to the World Bank/NGO Dialogue on Privatization, Washington.

Bourdieu, P. (1986). *Distinction*. London: Routledge.

Brockerhoff, M. and Brennan, E. (1998). 'The Poverty of Cities in Developing Regions', *Population and Development Review*, 24(1): 75–114.

Brustein, F. (1988*a*). 'Technologia y Servicios de Agua Potable y Saneamiento en America Latina', Centro de Estudios Urbanos y Regionales (CEUR), Buenos Aires, Paper presented at CIUDAGUA, Encuentro America-Latina, Europa sobre el acceso de la poblacion a los servicios de agua y saneamiento en los Ciudades de America Latina, Montevideo, 31 July–6 August 1988.

——(1988*b*). *Situación del Agua Potable y Saneamiento en el Area Metropolitana de Buenos Aires: Sintesis de Investigación*. Buenos Aires: Centro de Estudios Urbanos y Regionales (CEUR).

——(1988*c*). *Sistema Urbano y Servicios de Agua Potable y Saneaomiento: Informe de Avance de Investigacón.* Buenos Aires: Centro de Estudios Urbanos y Regionales (CEUR).

——(1990). 'Agua Potable y Crisis del Habitat en el Area Metropolitana de Buenos Aires: Discusión sobre Politicas Publicas', in *L'Eau, La Ville et le Développement—Water, City and Development—El Agua, La Ciudad y el Desarrollo.* Paris: Institut des Sciences et Techniques de l'Equipement et de l'Environnement pour le Développement (ISTED), 187–92.

——(1991). 'Decentralizaciòn de Servicios Urbanos, Experiencias Cooperativas de Saneamiento el Gran Buenos Aires', in Federacion Mundial de Ciudades Unidas (ed.), *Empresas Prestarias de Servicios de Agua y Alcantarillado y Usarios: ¿Pueden Tener y Desarrollar Proyectos Comunes? ¿Cuales? ¿Como?*. Quito: CIUDAD, 86–97.

Buck, Seifert, and Just, Consulting Engineers (1947). *Informe sobre el Diseño del Proyecto de Abastecimieno de Agua Potabe del Rio Daule para Servir a la Ciudad de Guayaquil, Ecuador*. Guayaquil: M. I. Consejo Cantonal de Guayaquil.

Burgess, R., Carmona, M., and Kolstee, J. (1997) (eds.). *The Challenge of Sustainable Cities.* London: Zed Books.

Burkett, P. (1999). *Marx and Nature: A Red and Green Perspective*. New York: St. Martin's Press.

Caballero, A. (1991). 'Proyectos UNUNCHIS, una Experiencia Alternativa de Participaciòn de la Poblaciòn Marginal en Cusco', in Federacion Mundial de Ciudades Unidas (ed.). *Empresas Prestarias de Servicios de Agua y Alcantarillado y Usarios: ¿Pueden Tener y Desarrollar Proyectos Comunes? ¿Cuales? ¿Como?*. Quito: CIUDAD, 178–80.

Cadme, J. and Morocho, D. (1980). 'El Deficit de la Vivienda en la Ciudad de Guayaquil', in *Explotaciòn y Miseria Urbana.* Guayaquil: Serie Investigaciones.

Calle, S. E. and Chang, L. L. (1976). *Estudio de la Legislaciòn de Desarrollo Urbano del Canton Guayaquil*. Ordenanza sustitutiva de la creación de la Empresa Municipal de Agua Potable de Guayaquil, Guayaquil: FEDESARROLLO.

Del Campo, E. (1980). 'Ecuador 1830–1980: Las Ciudades y la Politica', in *Politica y Sociedad. Ecuador 1830–1980*. Quito: Corporación Ed. Nacional.

Cans, R. (1994). *La Bataille de l'Eau*. Paris: Le Monde Editions.

Caravedo, M. B. (1986). *Autonomia y Efficiencia en las Empresas Publicas de Servicio. El caso SEDAPAL*. Lima: Fundacion Friedrich Ebert.

Carrión, F. (1986). 'Evoluciòn del Espacio Urbano Ecuatoriano', in F. Carrión (ed.), *El Proceso de Urbanizaciòn en el Ecuador del Siglo XVIII al Siglo XX*. Quito: Ed. El Conejo-CIUDAD, 147–8.

——(1992). 'Evolución del Espacio Urbano Ecuatoirano', in E. Ayala (ed.), *Nueva Historia del Ecuador, xii. Ensayos Generales I.* Quito: Corporación Editora Nacional/Editorial Grijalbo Ecuatoriana, 37–72.

Carrion, L. F. (1991). *Los Grupos Financieros en el Ecuador*. Quito: Centro de Educación Popular (CEDEP).

Castillo, G. (1988). 'Encuentro Guayaquil 2000', Colegio de Ingenieuros Civiles del Guayas—Centro de Actualización de Conocimientos, *Encuentro Guayaquil 2000*, 20–1 Oct.

Castree, N. (1995). 'The Nature of Produced Nature: Materiality and Knowledge Construction in Marxism', *Antipode*, 27(1): 12–48.

——(2001). 'Marxism, Capitalism, and the Production of Nature', in N. Castree and B. Braun (eds.), *Social Nature: Theory, Practice, and Politics*. Oxford: Blackwell, 189–207.

——and MacMillan, T. (2001). 'Actor-networks and the Reimagination of Nature', in N. Castree and B. Braun (eds.), *Social Nature: Theory, Practice, and Politics*. Oxford: Blackwell, 208–24.

Castro, E. (2002). 'Economic Arguments Underlying Current Programmes of Private Participation in Water and Sanitation Services in Developing Countries', Baseline document (draft), PRINWASS, Research Project, European Commission, INCO2 Research for Development, Contract PL ICA4-2001-10041. Oxford: School of Geography and the Environment, Oxford University.

——and Swyngedouw, E. (2000). *Metropolitan Cities and Sustainable Use of Water. Case Study: London*. Environment and Climate Programme Research Report, Framework IV Program Project. Brussels: European Commission, Directorate-General XII.

————and Kaïka, M. (2003). 'London: Structural Continuities and Institutional Change in Water Management', *European Planning Studies* (forthcoming).

Chadwick, E. (1842). *Report on the Sanitary Conditions of the Labouring Population of Great Britain.* London: B.P.P., vol.

——(1887). *The Health of Nations*. 2 vols., London: ed R. W. Richardson.

Chávez, F. M. (1944). *Crónicas de Guayaquil Antiguo.* Guayaquil: Imprenta y Talleres Municipales.

Chiriboga, M. (1980). *Jornaleros y Gran Proprietarios en 135 Años de Exportacion Cacaotera (1790–1925)*. Quito: Ed. CIESE, Consejo Provincial de Pichincha.

——(1988). 'Auge y Crisis de una Econommía Agroexportadora: el Período Cacaotera', in *Nueva Historia del Ecuador, ix. Cacao, Capitalismo y Revolucion Liberal.* Quito: Corporación Editora Nacional/Editorial Grijalo, 55–115.

Christmas, J. and De Roy, C. (1991). 'The Decade and Beyond: At a Glance', *Water International*, 16(3): 127–34.

CIUDAD (1988). 'Agua y Saneamiento en los Barrios Populares de Quito'. Quito (mimeographed).

Coing, H. and Montano, I. (1986). 'La Gestion de l'Eau Potable dans le Tiers-Monde'. *Les Annales de la Recherche Urbaine*, 30: 35–41.

Comité Ejecutivo Pro Agua Potable para el Suburbio Oeste (1989) 'Letter to Empresa Municipal de Agua Potable de Guayaquil, EMAP-G', Guayaquil, 5 July 1989 (mimeographed).

Comite Pro-Agua Potable y Progreso Comunitario (1990). *Historia de Una Gran Conquista*, Comité del Pueblo no. 1, Dpto. de Cultura y Difusion Popular. Quito: Universidad Central del Ecuador.

Conaghan, C. (1988). *Restructuring Domination*. Pittsburgh: University of Pittsburgh Press.

——and Espinal, R. (1990). 'Unlikely Transitions to Uncertain Regimes? Democracy without Compromise in the Dominican Republic and Ecuador'. *Journal of Latin American Studies*, 22: 553–74.

Coplain (1988). *Approvisonamiento de Agua Potable y Alcantarillado en las Ciudades Medias de Mexico*. Mexico City: Centro de la Vivienda y Estudios Urbanos, CENVI (mimeographed).

Corbin, A. (1994). *The Foul and the Fragrant*. London: Picador.

Corchuelo Rozo, A. (1981). *Capacidad de Pago de las Familias de Cali en Servicios Públicos*. Serie Monografias no. 5. Cali: Universidad del Valle.

Corporacion de Estudios y Publicaciones (1992). *Ley de Aguas y Reglamentos*. Quito.

Cortez, P. Y. (1992). 'Lucha Sindical y Popular en un Periodo de Transición', in *El Ecuador de la Postguerra*. Quito: Banco Central del Ecuador, 543–69.

Cotic, A. and Dascal, G. (1987). *Technologia y Saneamiento Urbano.* Informe de Investigación. Buenos Aires: Centro de Estudios Urbanos y Regionales (CEUR).

Craske, N. (1993). 'Women's Political Participation in Colonias Populares in Guadalajara, Mexico', in S. Radcliffe, and S. Westwood (eds.), *'Viva': Women and Popular Protest in Latin America*. London: Routledge, 112–35.

Crawford de Roberts, L. (1980). *El Ecuador en la Epoca Cacaotera*. Quito: Editorial Universitaria.

Crespo, C. (2002*a*). 'Water Privatisation Policies and Conflicts in Bolivia: The Water War in Cochabamba (1999–2000)', unpub. Ph.D. thesis, School of Planning, Oxford Brookes University, Oxford.

——(2002*b*). 'Structural Adjustment and Water Supply in Bolivia: Managing Diversity, Reproducing Inequality', in R. Zetter and R. White (eds.), *Planning in Cities: Sustainability and Growth in the Developing World*. London: ITDG Publishing, 189–202.

Cronon, W. (1991). *Nature's Metropolis: Chicago and the Great West*. New York: W. W. Norton.

Crow, B. (2002). 'Water: Gender and Material Inequalities in the Global South', Paper presented at the Conference of the American Association of Geographers, Los Angeles, March (mimeographed).

Cuervo, L. M., Bernal, P. I., Maldonado, M. M., Jaramillo, S., Alfonso, O., and Jaramillo, R. (1991). *Agua: Pasado y Presente: La Gestión del Servicio en Colombia*. Controversia 164–165, Bogota: Centro de Investigación y Educación Popular, 164–5.

Cueva, A. (1974). *El Proceso de Dominación Política en Ecuador*. Mexico: Editorial Diógenes.

——(1991). 'La Crisis de 1929–32: Un análisis', in *Las Crisis en el Ecuador: Los Treinta y Ochenta*, Bibliotheca de Ciencias Sociales, 33. Quito: Corporacion Editora Nacional, 61–77.

Davis, M. (1990). *City of Quartz: Excavating the Future in Los Angeles*. London: Verso.

——(1995). 'Los Angeles After the Storm: The Dialectic of Ordinary Disasters', *Antipode*, 37(3): 221–41.

——(1998). *Ecology of Fear: Los Angeles and the Imagination of Disaster.* New York: Metropolitan Books.

Debord, G. (1990). *Comments on the Society of the Spectacle*. London: Verso.

Deler, J. P. (1981). *Genèse de l'Espace Equatorien. Essai sur le Territoire et la Formation de l'Etat National.* Paris: Editions A.D.P.F.

Department for International Development (2000). *Strategies for Achieving the International Development Targets. Addressing the Water Crisis: Healthier and More Productive Lives for People.* Consultation document. London: DFID.

de Oliviera Filho, A. (2002). 'Brazil: Struggle Against the Privatization of Water', Paper presented at the International Water Conference, San José, Costa Rica, 8–10 July (mimeographed).

Desmarest, D. C. (1937). *An Adventure at Three Score And Ten*. Quito: Imprenta Nacional.

Dillon, L. N. (1985). *La Crisis Económico Financiera del Ecuador*. Guayaquil: Departamento de Publicaciones de la Faculdad de Ciencias Económicas de la Universidad de Guayaquil.

Dockes, P. (1969). *l'Espace dans la Pensée Economique du XVIe au XVIIIe Siècle.* Paris: Flammarion.

Donahue, J. M. and Johnston B. R. (1998) (eds.). *Water, Culture, and Power*. Washington: Island Press.

Egas, J. M. (1992). 'Genesis y Crisis del Populismo Clasico', in *El Ecuador de la Postguerra*. Quito: Editiones del Banco Central del Ecuador, 483–514.

El Universo. 'El Agua de Nuestras Cañerias', Guayaquil, 17 July 1979.

Eliot, T. S. (1969, first pub. 1922). 'The Waste Land', in *The Complete Poems and Plays.* London: Faber and Faber, 59–80.

EMAP-G (1980). *Plan Maestro de Agua Potable para Guayaquil y su Area de Influencia 1980–2000*. Guayaquil: Associacion Consultora Gilbert-Brown and Caldwell-Ribadeneira, 3 vols.

——(1988). *Ordenanza Reformada que Reglamento el Servicio de Agua Potable en el Canton Guayaquil y en la Jurisdicción Correspondiente*. Guayaquil: EMAP-G.

Enock, C. R. (1914). *Ecuador: Its Ancient and Modern History, Topography and Natural Resources, Industries and Social Development*. London and Leipzig: T. Fisher Unwin.

EPAP-G (1991*a*). *Programa de Control de Perdidas y Uso Eficiente del Agua de la Ciudad de Guayaquil y su Zona de Influencia*. Guayaquil: Direcciòn de Operaciones, Empresa Provincial de Agua Potable, Guayas.

——(1991*b*). 'Informe Ejecutivo 'Nuevo Systema de Abastecimiento de Agua Potable para la Ciudad de Guayaquil. Capacidad 10 m^3/Seg. comprende captacion, bombeo, impulsion, tratamiento y conduccion a Guayaquil'. Guayaquil: EPAP-G (mimeographed).

——(1992*a*). 'Memoria Descriptiva de Proyectos', Guayaquil: EPAP-G (mimeographed).

——(1992*b*). *Explotación de las Recursos Aquiferos al Este de la Ciudad de Guayaquil.* Guayaquil: Binnie & Partners/Overseas Development Administration.

Escobar, A. (1999). 'After Nature: Steps to an Antiessentialist Political Ecology'. *Current Anthropology*, 40(1): 1–30.

——(2001). 'Culture Sits in Places: Reflections on Globalism and Subaltern Strategies of Localization', *Political Geography*, 20: 139–74.

Espinoza, N. (1988). 'El Agua y Saneamiento en Lima y Callao', Paper presented at CIUDAGUA, Encuentro América-Latina—Europa sobre el acceso de la poblacion a los servicios de agua y saneamiento en los Ciudades de América Latina, Montevideo, 31 July–6 August 1988.

——and Oliden, J. C. (1988). 'Agua y Saneamiento en los Barrios Populares de Lima'. Lima: Centro de Investigacion, Documentacio y Asesoria Poblacional (CIDAP) (mimeographed).

Estache, A., Gomez-Lobo, A., and Leipziger, D. (2001). 'Utilities Privatization and the Poor: Lessons and Evidence from Latin America', *World Development*, 29(7): 1179–98.

Estrada, J. E. (1982). *Moneda y Bancos en el Ecuador*. Quito: Banco Central del Ecuador-Corporación Editora Nacional.

Estrada, J. Y. (1972). 'Evolución Urbana de Guayaquil', *Revista del Archivo Històrico del Guayas,* 1: 37–66.

——(1974). *La Fundación de Guayaquil.* Guayaquil: Publicaciones del Archivo Historico del Guayas.

Farrell, G. (1989). *La Investigación Economica en el Ecuador*. Antologia de las Ciencias Sociales, no. 4. Quito: ILDIS.

Faudry, D. (1987). *Les Problèmes d'Approvisionnement en Eau Potable et d'Assainissement dans les Villes d'Amérique Latine. Elements de Diagnostic et Etat de la Recherche*. Paris: Programme Interministériel REXCOOP.

——(1997). 'The Concession of Buenos Aires' Water Supply and Wastewater Services', in D. Lorrain (ed.), *Urban Water Management: French Experience Around the World.* Levallois Perret, Paris: Editions Hydrocom, 129–47.

Faudry, D., Ortiz, A., and Fandino, L. M. (1991). 'Los Servicios de l'Agua Potable y de Saneamiento Hidrico en las Ciudades Latinoamericanas', in Federacion Mundial de Ciudades Unidas (ed.), *Empresas Prestarias de Servicios de Agua y Alcantarillado y Usarios: Pueden Tener y Desarrollar Proyectos Comunes? ¿Cuales? ¿Como?*. Quito: CIUDAD, 78–80.

FEDEBAS (1993). *Por Una Vida Digna con Agua Potable para todos los Guaquileños.* Cuaderno de Trabajo. Guayaquil: FEDEBAS/FEDESO.

——(1996). *FEDEBAS-G Federación de Barrios Suburbanos de Guayaquil—¿Qué es Qué Hace y Hacia Dónde va la FEDEBAS-G?* Guayaquil: FEDEBAS/FEDOP/USAID.

Federación de Barrios Suburbanos de Guayaquil (FEDEBAS-G) (1992*a*). 'Analysis del Costo del Tank de Agua Transportada', Guayaquil, June/July 1992 (mimeographed).

——(1992*b*). 'Letter to Angel Duarte Valverde, Gobernador de la Provincia del Guayas', Guayaquil, 11 Sept. 1992 (mimeographed).

——(1993). 'Proyecto Borrador Propuesta de Trabajo Popular', Guayaquil, Sept. 1993 (mimeographed).

Financial Times Global Water Report (2000*a*). 'Bolivia: The Cochabamba Crisis', *Issue* 93: 1–3.

Financial Times Global Water Report (2000*b*). 'International Water's Rebuttal: Bolivia', *Issue* 94: 4–6.

Fitzsimmons, M. (1989). 'The Matter of Nature', *Antipode*, 21(2): 106–20.

Foster, J. B. (2000). *Marx's Ecology: Materialism and Nature*. New York: Monthly Review Press.

Frente de Usuarios del Guasmo (1990). 'Letter to Oswaldo Molestina Zavala, Gobernador de la Provincia del Guayas', Guayaquil, 20 Sept. 1990 (mimeographed).

——(1992). 'Estudio del Frente de Usuarios', Guayaquil, 11 Sept. 1992 (mimeographed).

Gaceta Municipal. Guayaquil, Nov. 1910.
——Guayaquil, 21 Feb. 1911.
——Guayaquil, 20 Mar. 1916.
Gandy, M. (1996). 'Crumbling Land: The Postmodernity Debate and the Analysis of Environmental Problems', *Progress in Human Geography*, 20(1): 23–40.
——(1997). 'The Making of a Regulatory Crisis: Restructuring New York City's Water Supply', *Transactions of the Institute of British Geographers*, 22: 338–58.
——(1999). 'The Paris Sewers and the Rationalization of Urban Space', *Transactions of the Insitute of British Geographers*, 24: 23–44.
——(2002). *Concrete and Clay: Reworking Nature in New York City*. Cambridge, Mass.: MIT Press.
Giddens, A. (1990). *Consequences of Modernity*. Cambridge: Polity Press.
Gilbert-Brown and Caldwell-Ribadeneira (1980). *Plan Maestro de Agua Potable y Alcantarillado*. Guayaquil: EMAP-G, 3 vols.
Gleick, P. (1993). *Water in Crisis*. Oxford: Oxford University Press.
——Wolff, G., Chalecki, E. L., and Reyes, R. (2002). *The New Economy of Water: The Risks and Benefits of Globalization and Privatization of Fresh Water.* Oakland, Calif.: Pacific Institute for studies in Development, Environment, and Security.
Godard, H. (1987). 'Quito, Guayaquil: Eje Central o Bicefalía', in *Geografia Basica del Ecuador, iii. El Espacio Urbano en el Ecuador*. Quito: Centro Ecuatoriano de Investigacion Geografica, 108–36.
——(1988). 'El Periodo Cacaotero en el Ecuador (1890–1925) y sus Repercusiones en Guayaquil', *Revista Geográfica*, 26: 139–72.
Goldenberg, N. R. (1981). 'l'Eau et l'Assainissement: Brasilia, l'Etat Est-il Producteur de Ségrégation Résidentielle et de Pénurie des Services?'. Thesis for Doctorat d'Etat, University of Paris XII, Val de Marne, October 1981.
González, V. A. (1988). *Guayaquil y su Historia: 1740–1987. Crònica y Reflexiones Epidemiològicas sobre la Fiebre Amarilla en la Ciudad.* Guayaquil: Universidad de Guayaquil.
Gottlieb, R. (1993). *Forcing the Spring: The Transformation of the American Environmental Movement.* Washington: Island Press.
——and Fitzsimmons, M. (1991). *Thirst for Growth*. Tucson: University of Arizona Press.
Goubert, J. P. (1989). *The Conquest of Water: The Advent of Health in Industrial Age.* Oxford: Blackwell and Cambridge: Polity Press.
Graham, S. and Marvin, S. (2001). *Splintering Urbanism: Networked Infrastructures, Technological Mobilities and the Urban Condition*. Routledge: London and New York.
Grijalva, W. M. (1990). 'La Economia Ecuatoriana de la Gran Recesión a la Crisis Bananero', in E. Ayala (ed.), *Nueva Historia del Ecuador*. Quito: Corporación Editoral Nacional, x. 37–69.
Grundman, R. (1991). *Marxism and Ecology*. Oxford: Clarendon Press.
Guerrero, A. (1980). *Los Oligarcas del Cacao: Ensayo sobre la Accumulacion Originaria en el Ecuador.* Quito: Ed. El Conejo.
Guy, S., Graham, S., and Marvin, S. (1996). 'Privatized Utilities and Regional Governance. The New Regional Managers?', *Regional Studies*, 30(8): 733–9.
————(1997). 'Splintering Networks: Cities and Technical Networks in 1990s Britain', *Urban Studies*, 34(2): 191–216.

Hall, D. (1999). 'The Water Multinationals'. Public Services International Research Unit, University of Greenwich. Occasional paper (mimeographed).
——and Lobina, E. (2002). 'Water Privatisation in Latin America, 2002', Public Services International Research Unit, University of Greenwich, Paper presented at the International Water Conference, San José, Costa Rica, 8–10 July 2000.
Hamerly, M. T. (1973). *Historia Social y Econòmica de la Antigua Provincia de Guayaquil 1763–1892*. Guayaquil: Publicaciones del Archivo Històrico del Guayas.
Haraway, D. (1991). *Simians, Cyborgs and Women: The Reinvention of Nature*. London: Free Association Books.
——(1997). *Modest-Witness@Second-Millennium.FemaleMan©-Meets_OncoMouse™*. London: Routledge.
Hardoy, A. and Schusterman, R. (1999). 'Las Privatizaciones de los Servicios de Agua Potable y Saneamiento y los Pobres Urbanos', *Medio Ambiente y Urbanización*, 15(54): 63–76.
Hardt, M. and Negri, A. (2000). *Empire*. Cambridge, Mass.: Harvard University Press.
Harvey, D. (1974). 'Class Monopoly Rent, Finance Capital and the Urban Revolution', *Regional Studies*, 8: 239–55.
——(1981). 'The Urban Process under Capitalism: A Framework for Analysis', in M. Dear and A. Scott (eds.), *Urbanization and Urban Planning in Capitalist Society*. London and New York: Methuen, 91–121.
——(1982). *Limits to Capital*. Oxford: Blackwell.
——(1985). *The Urbanization of Capital*. Oxford: Blackwell.
——(1996). *Justice, Nature and the Geography of Difference*. Oxford: Blackwell.
——(2002). *Spaces of Capital*. Edinburgh: Edinburgh University Press.
Hassan, J. (1998). *A History of Water in Modern England and Wales*. Manchester: Manchester University Press.
Haughton, G. (1996). 'Private Profits—Public Drought: The Creation of a Crisis in Water Management for West Yorkshire', *Sustainable Urban Development Working Paper Series*, 5, CUDEM, Leeds.
——and Hunter, C. (1994). *Sustainable Cities*. London: J. Kingsley.
Herrington, P. (1996). *Climate Change and the Demand for Water*. London: HMSO.
Hidalgo, P. (1932). 'Crecimiento de la Poblaciòn de Guayaquil y su Desarollo Urbano 1531–1931', *Revista Municipal*, March, 112–17.
Hundley, N. (Jr.) (1992). *The Great Thirst: Californians and Water, 1770s–1990s*. Berkeley and Los Angeles: University of California Press.
Hurtado, O. (1981). *El Poder Politico en el Ecuador*. Barcelona: Ed. Ariel.
——(1985). *Political Power in Ecuador*. Boulder, Col.: Westview Press.
Icaza, A. and Rodriguez, A. (1988). 'Agua Potable, Santiago de Chili', Centro de Investigaciones SUR (Santiago de Chili), Paper presented at CIUDAGUA, Encuentro América-Latina—Europa sobre el acceso de la poblacion a los servicios de agua y saneamiento en los Ciudades de América Latina, Montevideo, 31 July–6 August, 1988.
Illich, I. (1986). *H_2O and the Waters of Forgetfulness*. London: Marion Boyars.
INEC. *Censos Nacionales de Poblaciòn: Resultados Definitivos, 1950, 1962, 1974, 1982, 1990*. Quito: INEC.
Inter-American Development Bank (1997). 'IDB Approves $40 million to Improve Water, Sanitation in Guayaquil' (www.idb.org/exr/PRENSA/1997/cp16497e.htm—document 164/97—July 16, 1997).

Jacobi, P. (1995). *Environmental Problems Facing Urban Households in the City of São Paulo, Brazil.* Stockholm: Environment Institute.

Jaramillo, S. (1988). *La Evolucion del Suministro de los Servicios de Agua Potable y Desague en Bogota.* Bogota: Centro de Estudios sobre Desarollo Economico (CEDE)/Facultad de Economia, Universidad de los Andes.

Jessop, B. (1994). 'The Transition to Post-Fordism and the Schumpeterian Workfare State', in R. Burrows and B. Loader (eds.). *Towards a Post-Fordist Welfare State?* London: Routledge, London, 13–37.

Johns, M. (1992). 'The Urbanization of Peripheral Capitalism: Buenos Aires, 1880–1920', *International Journal of Urban and Regional Research*, 16(3): 352–74.

——(1993). 'The Antinomies of Ruling Class Culture: The Buenos Aires Elite, 1880–1910', *Journal of Historical Sociology*, 6(1): 74–101.

Jordan, S. and Wagner, F. (1993). 'Meeting Women's Needs and Priorities For Water and Sanitation in Cities', *Environment and Urbanisation*, 5(2): 135–45.

Kaïka, M. (1999). 'Modernity and the Urban Spaces of Produced Nature: The Politics and Culture of the Urbanisation of Water in Athens (1834–1999). unpub. D.Phil. thesis, School of Geography and the Environment, Oxford University.

——(2003*a*). 'Constructing Scarcity and Sensationalising Water Politics: 170 Days that Shook Athens', *Antipode* (forthcoming).

——(2003*b*). 'The Water Framework Directive: A New Directive for a Changing Social, Political and Economic European Framework', *European Planning Studies* (forthcoming).

——and Swyngedouw, E. (2000). 'Fetishizing the Modern City: The Phantasmagoria of Urban Technological Networks', *International Journal of Urban and Regional Research*, 24(1): 120–38.

Kalbermatten, J. (1980). *Appropriate Technology For Water Supply and Sanitation, iA. A Summary of Technical and Economic Options*. Washington: World Bank.

Kallis, G. and Coccossis, H. (2001). *Water for the City: Critical Issues and the Challenge of Sustainability*. Mytilini: Environmental Planning Laboratory, University of the Aegean.

Katz, C. (1998). 'Whose Nature, Whose Culture?: Private Productions of Space and the "Preservation" of Nature', in B. Braun and N. Castree (eds.), *Remaking Reality: Nature at the Millenium*. London: Routledge, 46–63.

Kazemir, B., Leinin, S., and Schaub, R. (2002). 'Privatization, Globalization, and Sustainability in the Water Industry'. Draft report, SUSTAINSERV Consulting, Swiss Institute for Environmental Science and Technology, Zurich.

Keil, R. (1994). 'A World of Natures: Space, Perception and the Construction of Urban Environments'. Paper read at Faculty of Environmental Studies, at York University, Toronto (mimeographed).

——(1995). 'The Environmental Problem in World Cities', in P. L. Knox and P. J. Taylor (eds.). *World Cities in a World System.* Cambridge: Cambridge University Press, 280–97.

——(1998). *Los Angeles: Globalization, Urbanization, and Social Struggles, World Cities Series*. New York: Wiley.

——(2000) (ed.). *Political Ecology: Global and Local*. London: Routledge.

——and Desfor, G. (1996). 'Making Local Environmental Policy in Los Angeles', *Cities*, 13(5): 303–13.

——and Graham, J. (1998). 'Reasserting Nature: Constructing Urban Environments after Fordism', in B. Braun and N. Castree N. (eds.), *Remaking Reality: Nature at the Millenium*. London: Routledge, 100–25.

Kirke, J. and Arthur, J. (1987). 'Water Supply Issues', in P. Richards and A. Thomson (eds.), *Basic Needs and the Urban Poor: The Provision of Communal Services: A Study Prepared for the ILO within the Framework of the World Employment Programme.* London: Croom Helm, 123–51.

Knaebel, G. and Leme, M. J. (1985). 'Les Escaliers Drainants de Salvador', *Les Annales de la Recherche Urbaine*, 25(Jan.), 61–81.

Komives, K. (2001). 'Designing Pro-Poor Water and Sewer Concessions: Early Lessons from Bolivia', *Water Policy*, 3(1): 61–80.

Larrea-Maldonado, L. (1982). 'Transnational Companies and Banana Exports from Ecuador, 1948–1972', *Northsouth*, 7(14): 3–42.

Laski, H. J., Jennings, W. I., and Robson, W. A. (1935) (eds.). *A Century of Municipal Progress 1835–1935*. London: Allen & Unwin.

Latour, B. (1993). *We Have Never Been Modern*. London: Harvester Wheatsheaf.

——(1998). 'To Modernise or to Ecologise? That is the Question', in B. Braun and N. Castree (eds.), *Remaking Reality: Nature at the Millenium*. London: Routledge, 221–42.

——(1999). *Politiques de la Nature: Comment Faire Entrer les Sciences en Démocratie*. Paris: La Découverte.

Lefebvre, H. (1991, first pub. 1974). *The Production of Space.* Oxford: Blackwell.

——(1995). *Introduction to Modernity*. London: Verso.

León, P. (1992). 'Dos Decadas de Producción, Comercio Exterior y Reordenamiento del Espacio', in *Ecuador de la Postguerra: Estudios en Homenaje a Guillermo Perez Chiriboga.* Quito: Banco Central del Ecuador, 93–116.

Lewontin, R. (1997). 'Genes, Environment, and Organisms', in R. B. Silvers (ed.). *Hidden Histories of Science.* London: Granta Books, 115–39.

Lipietz, A. (1986). 'New Tendencies in the International Division of Labor: Regimes of Accumulation and Modes of Regulation', in A. J. Scott and M. Storper (eds.), *Production, Work, Territory*. Boston: Allen & Unwin, 16–40.

——(1987). *Mirages and Miracles. The Crisis of Global Fordism*. London: Verso.

Lorrain, D. (1995) (ed.). *Gestions Urbaines de l'Eau*. Paris: Economica.

——(1997) (ed.). Urban Water Management. Levallois Perret: Editions Hydrocom.

Lykke, N. (1996). 'Between Monsters, Goddesses and Cyborgs: Feminist Confrontation with Science', in N. Lykke and R. Braidotti (eds.), *Between Monsters, Goddesses and Cyborgs*. London: Zed Books, 13–29.

Maiguashca, J. (1991). 'Los Sectores Sunalternos en los Años 30 y el Aparecimiento del Velasquismo', in *Las Crisis en el Ecuador: Los Treinta y Ochenta*, Bibliotheca de Ciencias Sociales, 33. Quito: Corporacion Editora Nacional, 79–93.

——(1992). 'La Cuestion Regional en la Historia Ecuatoriana (1830–1972)', in E. Ayala (ed.), *Nueva Historia del Ecuador, xii. Ensayos Generales I.* Quito: Corporación Editora Nacional/Editorial Grijalbo Ecuatoriana, 175–226.

——and North, L. (1991). 'Orígines y Significado del Velasquismo: Lucha de Clases y Participación Política en el Ecuador', in R. Quintero (ed.), *La Cuestion Regional y el Poder.* Quito: Corporacion Editora Nacional, 89–159.

Manrique, H. A. (1940). *Estudió del Aprovechamiento de las Aguas del Rio Daule para Abastecer de Agua Potable a la Ciudad de Guayaquil*, Guayaquil: Imprenta de la Universidad.

Marchán, C. (1988). 'Crisis Nacional, Aprovechamiento Regional y Discriminación Social de sus Efectos Econónomicos (1920–1927)', in C. Marchán (ed.), *Crisis y Cambios de las Economía Ecuatoriana en los Años Veinte*. Quito: Banco Central del Ecuador, 221–84.

——(1991). 'La Crisis del los Años Treinta: Differenciación Social de sus Efectos Económicos', in *Las Crisis en el Ecuador: Los Treinta y Ochenta*. Bibliotheca de Ciencias Sociales, 33. Quito: Corporacion Editora Nacional, 31–60.

Margat, J. (1992). 'Les Resources en Eau: Situation et Perspectives', Blue Plan no. 6. Paris: Economica.

Martinez, P. (1988). *Guayaquil Noviembre de 1922: Politica Oligárquica e Insurrección Popular*. Quito: Centro de Estudios y Difusion Popular (CEDIS).

Martinez-Allier, J. (1991). 'Ecology and the Poor: A Neglected Dimension of Latin American History', *Journal of Latin American Studies*, 23: 621–39.

Marvin, G. and Laurie, N. (1999). 'An Emerging Logic of Urban Water Management, Cochabamba, Bolivia', *Urban Studies*, 36(2): 341–57.

Marx, K. (1974, first pub. 1846). *The German Ideology* (ed. C. J. Arthur). London: Lawrence and Wishart.

——(1973, first pub. 1858). *Grundrisse*. New York: Vintage Books.

——(1971, first pub. 1867).*Capital*, vol. I. New York: Vintage Press.

Menendez-Carrión, A. (1986). *La Conquista del Voto. De Velasco a Roldós*. Quito: Corporación Editora Nacional/FLACSO.

——(1991). 'Expectativas, Clientelismo y Alianzas', in L. Verdesoto (ed.), *Gobierno y Política en el Ecuador Contemporáneo*. Quito: ILDIS, 407–69.

Merrifield, A. (1993). 'Place and Space: A Lefebvrian Reconciliation'. *Transactions of Institute of British Geographers*, 18: 516–31.

——and Swyngedouw, E. (1996) (eds.). *The Urbanization of Injustice*. London: Lawrence and Wishart.

Millward, B. (1991). 'Emergence of Gas and Water Monopolies in Nineteenth-Century Britain: Contested Markets and Public Control', in J. Foreman-Peck (ed.), *New Perspectives in Late Victorian Economy: Essays in Quantitative Economic History 1860–1914*. Cambridge: Cambridge University Press.

M. I. Municipio de Guayaquil (1933). *Ordenanza sobre Agua Potable*. Guayaquil.

——(1952). *Contrato para Estudios, Construcción y Administración de las Obras del Nuevo Aprovisionamiento de Agua Potable de Guayaquil*. Guayaquil.

Moncayo, C. (1974). 'Quito o Guayaquil? El Sistema Bicefálio Ecuatoriano'. Thèse de 3ième cycle, Université de Paris I—Sorbonne, Paris.

Montano, I. and Coing, H. (1985). *Le Service d'Eau Potable dans les Villes du Tiers-Monde: Modes de Gestion et d'Organisation*. Noisy-le-grand, Paris: Ecole Nationale des Ponts et Chaussées, Centre d'Enseignement et de Recherches, Techniques et Sociétés.

————(1991). 'Las Cooperativas de Agua en Argentina: Un Cuestionamiento a las Formas de Gestiòn Tradicionales', in Federacion Mundial de Ciudades Unidas (ed.), *Empresas Prestarias de Servicios de Agua y Alcantarillado y Usarios: ¿Pueden Tener y Desarrollar Proyectos Comunes? ¿Cuales? ¿Como?*. Quito: CIUDAD, 226–35.

Montesquieu, C. de Secondat, baron de (1973). *Lettres Persanes*. Edition établie et présentée par Jean Starobinski. Collection folio 475. Paris: Gallimard.

——(1995). *l'Esprit des Lois*. Paris: Nathan.

Montúfar, C. (1990). *1980–1990: La Crisis y el Desarollo Social en El Ecuador*. Quito: Editorial El Conejo/UNICEF.
Moore, C. W. and Lidz, J. (1994). *Water and Architecture*. London: Thames and Hudson.
Moreno Mendoza, C. (1998). 'Agua que no has de Beber', *Vistazo*, 740, 4 June 1998.
Moser, C. (1987). 'Mobilization is Women's Work: Struggles for Infrastructure in Guayaquil, Ecuador', in C. Moser (ed.), *Women, Human Settlements and Housing*. London: Tavistock, 166–94.
Moulaert, F. and Swyngedouw, E. (1987). 'A Regulation Approach to the Geography of the Flexible Production System', *Environment and Planning D: Space and Society*, 7: 327–45.
Multilateral Investment Guarantee Agency (2002). 'Project "homes in" on Water Services for Ecuador's Poor' (www.miga.org/screens/news/feature/water.htm—5 Sept. 2002.).
Mumford, L. (1961). *The City in History. Its Origins, its Transformation and its Prospects*. London: Secker and Warburg.
Muños Vicuña, E. (1978). *El 15 de Noviembre de 1922*. Guayaquil: Departamento de Publicaciónes de la Faculdad de Ciencias Económicas de la Universidad de Guayaquil.
——(1987). *La Guerra Civil Ecuatoriana en 1895*. Guayaquil: Departamento de Publicaciónes de la Faculdad de Ciencias Económicas de la Universidad de Guayaquil.
——and Vicuña Izquierdo, L. (1978). *Historia del Movimineto Obrero del Ecuador: Resumen*. Guayaquil: Facultad de Economía, Universidad de Guayaquil.
Nevarez, L. (1996). 'Just Wait Until There's a Drought: Mediating Environmental Crises for Urban Growth', *Antipode*, 28(3): 246–72.
New Delhi Statement (1991). 'Some for All Rather than More for Some', *Water International*, 16(3): 115–18.
Niemczynowicz, J. (1991). 'Water Management and Urban Development', *Impact of Science on Society*, 116: 131–47.
——(1992). 'Water Management and Urban Development: A Call for Realistic Alternatives for the Future', *Environment and Development*, 166: 131–47.
Noorgaard, R. (1994). *Development Betrayed*. London: Routledge.
Nurse, C. (1989). 'Ecuador', in J. Carriere, N. Haworth, and J. Roddick (eds.), *The State, Industrial Relations and the Labour Movement in Latin America*. London: Macmillan, i. 99–127.
O'Connor, M. (1994). 'Codependency and Indeterminacy: A Critique of the Theory of Production', in M. O'Connor (ed.), *Is Capitalism Sustainable? Political Economy and the Politics of Ecology*. New York: The Guilford Press, 53–75.
O'Connor, J. (1998) (ed.). *Natural Causes: Essays in Ecological Marxism*. New York: Guilford Press.
Olaya, N. (1991). *La Crisis de la Basura en Guayaquil*. Series 'Guayaquil Futuro'. Quito: Instituto Latinoamericano de Investigaciones Sociales (ILDIS).
——and Villavicencio, G. (n.d.). 'Vivir en el Agua y Morirse de Sed: El Caso del Abastecimiento del Agua Potable en Guayaquil', unpub. paper, Corporacion de Estudios Regionales, Guayaquil (CER-G), Guayaquil.
————(1990). 'El Abastecimiento de Agua Potable en las Areas Populares de Guayaquil: Historia de una Imposible Cooperacion Participativa', Quito: CIUDAD, 22–6 Jan.

Olman, B. (1993). *Dialectical Investigations*. London: Routledge.

Pahl, R. (1996). *Whose City?* Harmondsworth: Penguin.

Peet, R. and Watts, M. (1996) (eds.). *Liberation Ecologies*. London: Routledge.

Perez Pimentel, R. (1987). *Nuestro Guayaquil Antiguo*. Guayaquil: Ed. El Sol.

Petrella, R. (1998). *La Manifeste de l'Eau*. Brussels: Editions Labor.

Pineo, R. F. (1988). 'Reinterpreting Labor Militancy: The Collapse of the Cacao Economy and the General Strike of 1922 in Guayaquil, Ecuador'. *Hispanic American Historical Review*, 68(4): 707–36.

——(1990). 'Misery and Death in the Pearl of the Pacific: Health Care in Guayaquil, Ecuador, 1870–1925'. *Hispanic American Historical Review*, 70(4): 609–37.

——(1996). *Social and Economic Reform in Ecuador: Life and Work in Guayaquil*. Gainsville, Fla.: University Press of Florida.

Postel, S. (1992). *The Last Oasis*. London: Earthscan.

PREDAM (1976). *Análisis Socio-Econòmico Preliminar de las Areas de Concentraciòn de la Pobreza de la Ciudad de Guayaquil*. Quito: Junta Nacional de Planificación.

Quintero, R. (1980). *El Mito del Populismo en el Ecuador*. Quito: Facultad Latino Americano de Ciencias Sociales (FLACSO).

——(1991). 'El Mito del "Populismo Velasquista" y la Consumacion del Pacto Oligarquico', in L. Verdesoto (ed.), *Gobierno y Política en el Ecuador Contemporáneo*. Quito: ILDIS, 181–217.

——and Silva, E. (1991). 'Región y Representación Política en el Ecuador', in R. Quintero (ed.), *La Cuestion Regional y el Poder*. Quito: Corporacion Editora Nacional, 29–87.

Radcliffe, S. and Westwood, S. (1993) (eds.). *'Viva': Women and Popular Protest in Latin America*. London: Routledge.

Reisner, M. (1986). *Cadillac Desert. The American West and its Disappearing Water*. Harmondsworth: Penguin.

Revista Municipal (1936). Guayaquil, July–Aug. 1936, nos. 29–30.

——(1938). Guayaquil, 1 Feb. 1938.

——(1939). 'Proyecto de Obras Públicas y su financimiento', Guayaquil, June 1939.

——(1948). 'Memorial del M. I. Concejo Cantonal de Guayaquil, al H. Congreso Nacional, reclamando la Ejecución de las Obras de Saneamiento de la Ciudad, mediante la restitución del producto de los fondos especiales que fueron destinados para tal finalidad, desde 1899', Guayaquil, Oct.–Dec. 1948.

Ridgley, M. A. (1989). 'Services in a Colombian Shantytown: Speculations on the Limits of Collective Self-Help', *Yearbook, Conference of Latin Americanist Geographers*, 15: 59–69.

Rindisbacher, H. J. (1992). *The Smell of Books*. Ann Arbor: University of Michigan Press.

Robelino, I. B. (1976). *El Sindicalismo en El Ecuador*. Quito: INEDES.

Roberts, R. S. and Emel, J. (1992). 'Uneven Development and the Tragedy of the Commons: Competing Images for Nature-Society Analysis'. *Economic Geography*, 68(3): 249–71.

Rodriguez, J. (1987). 'El Proceso de Urbanizacion en la Costa Ecuatoriana', in *Geografia Basica del Ecuador' iii. El Espacio Urbano en el Ecuador*. Quito: Centro Ecuatoriano de Investigacion Geografica, 158–70.

Rodríguez, L. A. (1984). *The Search for Public Policy: Regional Politics and Government Finances in Ecuador, 1830–1940*. Los Angeles and Berkeley: University of California Press.

Rogers, T. (1992*a*). 'Neighborhood Movements in Latin America: The Case of FEDEBAS, A United Neighborhood Front in Guayaquil, Ecuador', unpub. research paper, Latin American Studies and Community and Regional Planning, University of New Mexico, Albuquerque.

——(1992*b*). *Low-Income Housing in Guayaquil, Ecuador*. Unfinished manuscript, Latin American Studies and Community and Regional Planning, University of New Mexico, Albuquerque.

Rojas, M. and Villavicencio, G. (1988). *El Proceso Urbano de Guayaquil*. Quito/Guayaquil: Instituto Latinoamericano de Investigaciones Sociales (ILDIS) and Centro de Estudios Regionales de Guayaquil (CER-G).

Rosenfeld, A. (1991). 'Tarifas y Participación', in Federación Mundial de Ciudades Unidas (ed.), *Empresas Prestarias de Servicios de Agua y Alcantarillado y Usarios: ¿Pueden Tener y Desarrollar Proyectos Comunes? ¿Cuales? ¿Como?*. Quito: CIUDAD, 187–200.

Roy, A. (2001). *Power Politics*. Cambridge, Mass.: South End Press.

Saenz, R. (1988). 'El Agua y el Saneamiento Urbano en la Region de las Americas', Programma de Salud Ambiental, Organizacion Panamericana del Salud, Washington (mimeographed).

Satterthwaite, D. (1999). *The Earthscan Reader in Sustainable Cities.* London: Earthscan.

Savedoff, W. and Spiller, P. (1999). *Spilled Water: Institutional Commitment in the Provision of Water Services*. Washington: Interamerican Development Bank (IDB).

Schama, S. (1995). *Landscape & Memory*. London: HarperCollins.

Scheers, J. (1991). 'Invasiones en Guayaquil of Hyperurbanisatie tussen Esteros en Rio Guayas'. unpub. M.Sc. Dissertation, Instituut voor Stedebouw en Ruimtelijke Ordening, Catholic University, Leuven.

——(2002). 'Het Aroma van de Stad. Koffie, Migratie en Stedelijke Onwikkeling in Guayaquil, Ecuador'. unpubl. Ph.D. dissertation. Department of Sociology, Catholic University, Leuven.

Schmidt, A. (1971). *The Concept of Nature in Marx*. London: New Left Books.

Schodt, D. W. (1987). *Ecuador: An Andean Enigma*. Boulder, Col.: Westview.

Scobie, A. (1986). 'Seums, Sanitation, and Mortality in the Roman World.' *Klio*, 68: 399–433.

Sennett, R. (1994). *Flesh and Stone*. London: Faber and Faber.

Servicios Sociales del Suburbio (1976). *El Otro Guayaquil*. Guayaquil.

Shiklomanov, I. (1990). 'Global Water Resources'. *Nature and Resources*, 26(3): 34–43.

——(1997) (ed.). *Assessment of Water Resources and Water Availability in the World.* St Petersburg: State Hydrological Institute.

Shiva, V. (2002). *Privatization, Pollution, and Profit.* Cambridge, Mass.: South End Press.

Smith, N. (1984). *Uneven Development: Nature, Capital and the Production of Space.* Oxford: Blackwell.

——(1996). 'The Production of Nature', in G. Robertson, M. Mash, L. Tickner, J. Bird, B. Curtis, and T. Putnam (eds.), *FutureNatural: Nature/Science/Culture.* London: Routledge, 35–54.

——(1998). 'Antinomies of Space and Nature in Henri Lefebvre's "The Production of Space"', in A. Light and J. M. Smith (eds.), *Philosophy and Geography, ii. The Production of Public Space.* London and New York: Rowman and Littlefield, 49–70.

Solo, T. M. (2001). 'Independent Water Entrepreneurs in Latin America: The Other Private Sector in Water Services', *Business Briefing: World Infrastructure and Development*. Official publication of the World Bank Infrastructure Forum (2–11 May 2001), Washington. (www.wmrc.com/businessbriefing/pdf/infrastructure2001/reference/37.pdf)

Stauffer, J. (1998). *The Water Crisis*. London: Earthscan.

Stephen, L. (1992). 'Women in Mexico's Popular Movements: Survival Strategies Against Ecological and Economic Impoverishment', *Latin American Perspectives*, 19(1): 73–96.

Suarez-Torres, J., Lopez-Paredes, D., Uggen, J. F., and Crawford, S. (1997). 'Development, Environment, and Health Care in Crisis: The Case of Ecuador'. *Latin American Perspectives*, 24(3): 83–103.

Suskind, P. (1987). *Perfume*. London: Penguin.

Swyngedouw, E. (1994). 'Watervoorziening Guasmo en Suburbio, Guayaquil: Aankoop van 5 Tankwagens en Organizatie van de Waterverkoop en-verdeling in (delen van) Guasmo en Suburbio', VLIR-Project, IPUR/ISRO, Guayaquil/Leuven (mimeographed).

——(1995*a*). *El Problema del Abastecimiento del Agua Potable en Guayaquil*. Quito: ILDIS.

——(1995*b*). 'The Contradictions of Urban Water Provision'. *Third World Planning Review*, 17(4): 387–405.

——(1996*a*). *Water and the Politics of Economic Development in Southern Spain Final Report*. Directorate-General XII, Commission of the European Communities, Brussels, (contract n° JPBSMBICT961054).

——(1996*b*). 'The City as a Hybrid: On Nature, Society and Cyborg Urbanisation', *Capitalism, Nature, Socialism*, 7(1), Issue 25 (Mar.), 65–80.

——(1996*c*). 'Producing Futures: Global Finance as a Geographical Project', in P. W. Daniels, and W. F. Lever (eds.). *The Global Economy in Transition*. Oxford and London: Longman, 135–63.

——(1997). 'Power, Nature and the City. The Conquest of Water and The Political Ecology of Urbanization', *Environment and Planning A*, 29(2): 311–32.

——(1998). 'Homing In and Spacing Out: Re-Configuring Scale', in H. Gebhardt, G. Heinritz, and R. Weissner (eds.), *Europa im Globalisieringsprozess von Wirtschaft und Gesellschaft*. Stuttgart: Franz Steiner Verlag, 81–100.

——(1999). 'Modernity and Hybridity: Nature, *Regeneracionismo*, and the Production of the Spanish Waterscape, 1890–1930', *Annals of the Association of American Geographers*, 89(3): 443–65.

——and Bovarnick, A. (1995). *Flujos de Poder: l'Abastecimiento de Agua Potable en Guayaquil*. Quito: Institute Latinoamericano de Investigaciones Sociales (ILDIS).

——and Kaïka, M. (2000). 'The Environment of the City or—the Urbanization of Nature', in G. Bridge and S. Watson (eds.), *Companion to Urban Studies*. Oxford: Blackwell, 567–80.

——Page, B. and Kaïka, M. (2002). 'Sustainability Policy Innovation in a Multi-Level Context: Crosscutting Issues in the Water Sector', in H. Heinelt, P. Getimis, P. G. Kafkalas, R. Smith, and E. Swyngedouw (eds.), *Participatory Governance in Multilevel Context: Concepts and Experience*. Opladen: Leske and Budrich, 107–31.

Teich, M. (1982). 'Circulation, Transformation, Conservation of Matter and the Balancing of the Biological World in the Eighteenth Century', *Ambix*, 29: 17–28.

Thomas, C. and Howlett, D. (1993) (eds.). *Resource Politics: Freshwater and Regional Relations*, Buckingham/Philadelphia: Open University Press.

Tobar, L. (1992). 'Panoramo Agrario desde 1948', in *Ecuador de la Postguera, Estudios en Homenaje a Guillermo Perez Chiriboga*. Quito: Banco Central del Ecuador, 227–61.

Trujillo, J. (1992). 'Expansión de la Frontera Agricola', in *El Ecuador de la Postguerra, Estudios en Homenaje a Guillermo Perez Chiriboga*. Quito: Banco Central del Ecuador, 201–26.

Uclés, M. L. (1991). 'La Gestión de los Servicios de Agua en Algunas Ciudades Centroamericanos', in Federacion Mundial de Ciudades Unidas (ed.), *Empresas Prestarias de Servicios de Agua y Alcantarillado y Usarios: ¿Pueden Tener y Desarrollar Proyectos Comunes? ¿Cuales? ¿Como?*. Quito: CIUDAD, 277–81.

United Nations Centre for Human Settlements (UNCHS) (1985). *Infrastructure Technology and Design*. Nairobi: United Nations Centre for Human Settlements.

——(1991). 'Urbanization: Water Supply and Sanitation Sector Challenges', Paper presented at the 'Water Supply and Sanitation Collaborative Council Global Forum', Oslo, 18–20 Sept. 1991.

——(2001). *Cities in a Globalizing World: Global Report on Human Settlements*. United Nations Centre for Human Settlements. London: Earthscan.

Unda, M. (1986). 'La Organicaciòn Barrial Entre la Democracia y la Crisis', in CACSO/ILDIS (eds.), *Movimientos Sociales en el Ecuador*. Quito: CACSO/ILDIS, 219–79.

UNEP (2002). *Global Environmental Outlook 3*. New York: United Nations Environment Programme.

Universidad Catolica (1982). 'Perspectivas del Desarrollo Urbano de Guayaquil: Documento Base', Seminario Internacional de Urbanismo, Universidad Catolica, Guayaquil (mimeographed).

Vásconez, M. (1988*a*). 'Agua y Saneamiento en los Barrios Populares de Quito', CIUDAD, Quito (mimeographed).

——(1988*b*). 'Segregacion Urbana, Necessidades de la Poblacion y Formas de Abastecimiento de Agua y Saneamiento', Paper presented at CIUDAGUA, Encuentro América-Latina–Europa sobre el acceso de la poblacion a los servicios de agua y saneamiento en los Ciudades de América Latina, Montevideo, 31 July–6 Aug.

——(1991). 'Participación Popular en la Busqueda de Solución a las Carencias de Agua Potable y Alcantarillado', in Federacion Mundial de Ciudades Unidas (ed.), *Empresas Prestarias de Servicios de Agua y Alcantarillado y Usuarios: ¿Pueden Tener y Desarrollar Proyectos Comunes? ¿Cuales? ¿Como?*, Quito: CIUDAD, 47–56.

——(1992). *Agua y Sociedad*. Quito: CIUDAD.

Vernes, J. (1994). *Paris dans le XXième Siècle*. Paris: Hachette.

Vigarello, G. (1988). *Concepts of Cleanliness: Changing Attitudes in France since the Middle Ages*. Cambridge: Cambridge University Press.

Villavicencio, G. (1988). 'Las ofertas electorales y los limites del Clientelismo', in *Ecuador 88*. Quito: Editorial El Conejo, 11–33.

——(1990). 'El Desfase de un Proceso Urbano: El Caso de Guayaquil', in CORDES (ed.), *Guayaquil, Relidades y Desafios*, Quito: Corporación de Estudios para el Desarollo, 29–87.

——(1992). 'Historia del Agua Potable en Guayaquil'. Centro de Estudios Regionales de Guayaquil (CER-G). Internal document (mimeographed).

Villavicencio, G., Rojas, M., and Olaya, N. (1988). 'Población sin Redes de Agua y Saneamiento: El Caso de Guayaquil y sus Barrios Populares (Suburbio del Suroeste y el Guasmo)', Paper presented at conference *CIUDAGUA ANDINA*, Montevideo, 31 July–6 Aug. (mimeographed).

Vlaams Internationaal Centrum (1994). 'Watervoorziening in de Barrios Guasmo en Suburbio, Guayaquil, Ecuador'. Brussels, Vlaams Internationaal Centrum (mimeographed).

Ward, C. (1997). *Reflected in Water. A Crisis of Social Responsibility*. London and Washington: Cassell.

Watkins, M. (1991). 'Proposal for Technical and Financial Cooperation: Development of the Aquifer Resources East of the City of Guayaquil'. Binnie & Partners, Consulting Engineers, Interim draft report, London, Nov. 1991.

Whatmore, S. (2002). *Hybrid Geographies*. London: Routledge.

Williams, R. (1985, first pub. 1973). *The Country and the City.* London: Hogarth Press.

Wittfogel, K. (1957). *Oriental Despotism: A Comparative Study of Total Power.* New Haven: Yale University Press.

Wolf, T. (1892). *Geografia y Geologia del Ecuador*. Quito and Leipzig: Supremo Gobierno de la Republica.

Wolman, A. (1965). 'The Metabolism of Cities', *Scientific American*, 213: 179–90.

World Bank (1976). *Village Water Supply.* Washington: World Bank.

——(1987). *World Bank Atlas*. Oxford: Oxford University Press.

——(1989). *Urban Strategy Paper*. Washington: World Bank.

——(1992). *World Development Report 1992: Development and the Environment*. Oxford: Oxford University Press.

——(1998). *World Resources 1998–99*. Oxford: Oxford University Press.

——(2000). *World Development Report 2000/2001.* Oxford: Oxford University Press.

Worster, D. (1985). *Rivers of Empire: Water, Aridity and the Growth of the American West*. New York: Pantheon.

Wright, L. (1960). *Clean and Decent: The Fascinating History of the Bathroom and the Water Closets*. Routledge and Kegan Paul: London.

Ycaza, P. (1992). 'Lucha Sindical y Popular en un Periodo de Transicion (1948–1970)', in *El Ecuador de la Postguerra.* Quito: Banco Central del Ecuador, Quito, 201–26.

Yepes, G. (1999). 'Do Cross-Subsidies Help the Poor to Benefit from Water and Wastewater Services? Lessons from Guayaquil'. Working Paper Series. Water and Sanitation Program. Wahington: United Nations Development Programme/World Bank.

Zavala, I. (1990). 'Nouvelle Politique de la Banque Mondiale', in *L'Eau, La Ville et le Développement—Water, City and Development—El Agua, La Ciudad y el Desarrollo*. Paris: Institut des Sciences et Techniques de l'Equipement et de l'Environnement pour le Développement (ISTED), 407–12.

Zolezzi, M. and Calderon, J. (1985). *Vivienda Popular: Autoconstruction y Lucha por el Agua*. Lima: Centro de Estudios y Promocion del Desarrollo (DESCO).

Zukin, S. (1991). *Landscapes of Power*. Berkeley: University of California Press.

INDEX